Serious Managers Guide to AI Strategy and Road Mapping

Step-by-Step Overview of Implementation for an Agentic-based Workforce, including Challenges and Recommendations

(Hands-on workbooks included)

Claude Louis-Charles, PhD

Brittanee Charles

Cybersoft Publishing LLC

Fort Washington, MD 20744

First Edition March 2026

Acknowledgment

I (Claude) want to express my gratitude to my peers who have, unknowingly, served as my sounding board. I truly appreciated their invaluable patience and feedback for my endless series of questions and ideas. This includes, but is not limited to, Matthew Wilson, Anthony Egbuniwe, and Patrick Munis. I am endlessly thankful to my loving wife, Tracy, who has supported me for over 30 years, and my children, JC and Jordan, who have been cheering me on no matter how many books I choose to write. Their unwavering belief in me has kept my spirits and motivation high throughout this lengthy process.

I (Claude) want to express my gratitude to my very spoiled dogs, Ozzy and Jazz, for providing entertainment and emotional support during the many late nights of research.

I (Brittanee) am deeply grateful to my husband, Chris, and my son (Isaiah), whose love, patience, and unwavering belief carried me through every late night and early morning of this journey. Their encouragement gave me the confidence to keep pushing, their sacrifices created the space for me to write, and their presence reminded me of what truly matters. This book exists because of their strength, their support, and the joy they bring into my life every day.

Table of Contents

Introduction

Artificial intelligence has shifted from experimental use to a core element of business strategy. It is now a top priority, vital to competitive advantage, and essential in industries such as healthcare and the public sector. Yet despite the urgency, most organizations still struggle to answer the most fundamental questions:

Where should we start?

What should we build?

How do we ensure AI is safe, responsible, and aligned with our mission?

How do we avoid wasting time, money, and credibility?

These questions aren't technical. They're managerial, and they're exactly why this book exists.

The Leadership Gap in AI Adoption

AI is advancing at a pace that outstrips most organizations' ability to absorb it. Technical teams understand the models. Executives understand the pressure. But the middle—the managers responsible for translating ambition into action—often lack clear, practical guidance.

You're expected to make decisions about AI investments, prioritize use cases, manage risk, and lead teams through change. Yet most resources fall into one of two extremes:

- **Deeply technical content** that assumes you're training models yourself

- **High-level thought leadership** that sounds inspiring but offers no actionable steps

Managers need something different: a grounded, operational playbook that connects strategy to execution.

This book fills that gap.

AI Strategy Is Not Optional

Organizations that treat AI as a collection of disconnected experiments quickly find themselves in trouble. They accumulate technical debt, duplicate efforts, and create inconsistent standards. They chase hype instead of outcomes. They deploy tools without understanding the risks. They burn out teams and lose trust.

In contrast, organizations with a clear AI strategy:

- Prioritize high-value, mission-aligned use cases
- Build responsibly and avoid unnecessary risk
- Scale efficiently instead of reinventing the wheel
- Communicate clearly and bring people along
- Deliver measurable impact

AI strategy is not a luxury. It is the foundation for everything that follows.

Why does this book start with Roadmapping?

Many leaders jump straight into tools, vendors, or pilots. But without a roadmap, even the best technology fails. A roadmap forces clarity. It forces sequencing. It forces tradeoffs. It forces alignment.

A good roadmap answers:

• What are we trying to achieve?
• What capabilities do we need?
• What should we build first?
• What risks must we take?
• How do we measure progress?

This book gives you a 90-day roadmap framework that has been tested across healthcare, public sector, and enterprise environments. It is simple enough to use immediately and robust enough to guide long-term transformation.

AI Strategy Is About People, Not Just Tech.

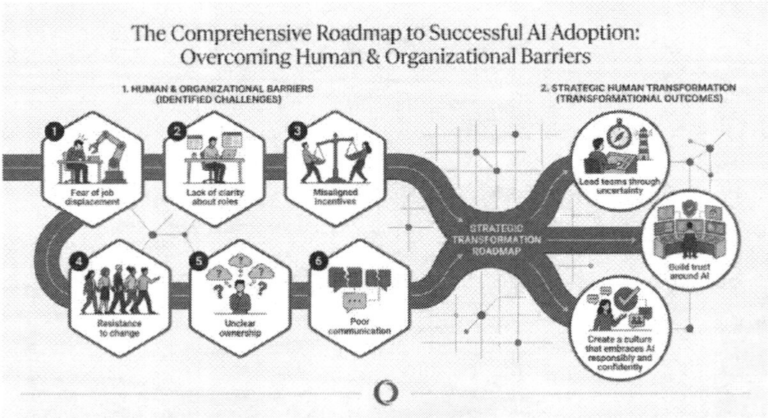

Figure 1: Real Barrier to AI Adoption

The biggest barriers to AI adoption are rarely technical. They are cultural, organizational, and operational:

• Fear of job displacement
• Lack of clarity about roles
• Misaligned incentives
• Resistance to change
• Unclear ownership

• Poor communication

This book treats AI as a human transformation as much as a technological one. You'll learn how to lead teams through uncertainty, build trust, and create a culture that embraces AI responsibly and confidently.

A Practical, Manager-Focused Approach

This book is intentionally pragmatic. You won't find mathematical formulas, model architectures, or deep technical dives. Instead, you'll find:

- Frameworks you can use in meetings tomorrow
- Templates for strategy, governance, and roadmapping
- Real-world examples from high-stakes environments
- Clear explanations of what matters—and what doesn't
- Guidance on how to partner effectively with technical teams

You don't need to be a data scientist to lead AI initiatives. You need clarity, structure, and the ability to connect mission, people, and technology. That's what this book gives you.

A Commitment to Responsible AI

An AI strategy without responsible AI is incomplete. Every decision about building, deploying it, and measuring success carries implications for safety, fairness, privacy, and trust. This book integrates responsible AI throughout, not as a compliance exercise but as a strategic advantage.

Responsible AI is not about slowing down. It's about scaling safely, sustainably, and credibly.

What You Will Learn

By the end of this book, you will know how to:

- Build an AI strategy grounded in mission and outcomes
- Assess your organization's readiness across data, people, processes, and culture
- Identify and prioritize high-value use cases
- Design an AI operating model that fits your organization
- Build a 90-day roadmap that drives momentum
- Run pilots that lead to real adoption—not dead ends
- Scale AI responsibly across the enterprise
- Lead teams through change and build a culture of AI adoption

This is the guide I wish every manager had before being asked to "figure out AI."

Importance of Each Chapter's Workbooks

Working through the workbook at the end of each chapter is essential because it turns the book from something you understand into something you can use.

If you are reading this via an e-book, you are encouraged to write down the Exercises on a separate piece of paper.

Each chapter gives you concepts, frameworks, and mental models for serious AI strategy, but the workbook forces you to translate those ideas into the realities of your organization, your mission, your workflows, your data, your risks, your constraints, and your opportunities. By slowing down to map concepts onto your environment, you build the muscle memory of strategic thinking, not just the vocabulary. The workbook also creates a cumulative record of decisions, assumptions, and priorities that becomes your living AI strategy artifact—something you can share with peers, refine with stakeholders, and revisit as conditions change. In a field where clarity, alignment, and iteration determine success, the workbook is where theory becomes leadership. Using the workbox allows you the opportunity to:

"Fail on Paper and Adjust on Paper"

A Final Word Before We Begin

AI is not just another technology wave. It is a fundamental shift in how organizations operate, make decisions, and achieve their missions. The leaders who succeed in this era will be those who approach AI with clarity, responsibility, and strategic intent.

You don't need to know everything about AI. You need to know how to lead.

Let's begin.

Part I – Understanding the AI Landscape

Part I, *Understanding the AI Landscape*, lays the groundwork every serious manager needs before making decisions, approving budgets, or evaluating proposals. This section demystifies the technology by explaining what today's AI systems are, how they work at a high level, and why they behave the way they do. It gives leaders a clear vocabulary for discussing AI with technical teams and vendors, and it builds the strategic intuition required to separate meaningful capability from marketing noise. By the end of Part I, managers gain a realistic, evidence-based view of the current AI ecosystem—its strengths, its blind spots, and its trajectory—so they can lead with confidence rather than react to hype.

Just as importantly, this section reframes AI as an operational tool rather than a futuristic abstraction. It shows how different categories of AI capabilities map to real workflows, how the limits of current models create risk, and how to evaluate feasibility without being a data scientist. Part I equips managers to ask sharper questions, identify red flags early, and recognize when a use case is technically plausible rather than wishful thinking. With this foundation, the rest of the guide becomes far more actionable: strategy, mission alignment, workflow analysis, and implementation all make sense only when grounded in a clear understanding of the landscape itself.

1 The AI Moment: Why Strategy Matters Now

1.1 The Pressure to "Do Something with AI."

Opening Scenario: You're in a leadership meeting. Someone mentions that a competitor just launched an AI-powered service. Another leader says their team is already experimenting with a chatbot. A vendor claims they can automate 40% of your workflows. Your CIO warns that without guardrails, AI could create new risks. Your CEO wants a plan fast.

Everyone agrees that AI is important. No one agrees on what to do next.

This is the AI moment managers are living in: high expectations, high uncertainty, and high stakes. The pressure to act is real. The risk of acting without clarity is even greater.

This book begins here—at the intersection of urgency and ambiguity where strategy matters most.

1.1 Why AI Strategy Matters More Than Ever

AI is no longer a future technology. It is a present-day operational force. Organizations are deploying AI to streamline workflows, improve decision-making, enhance customer experience, and reduce administrative burden. In healthcare and the public sector, AI is reshaping how clinicians, administrators, and mission teams work.

But here's the truth most leaders won't say out loud:

1. AI is easy to get started with, but hard to scale.

Most organizations begin with enthusiasm and end with fragmentation:

- Dozens of disconnected pilots
- Competing vendor solutions
- Shadow AI built without oversight
- Teams are unsure how to use new tools
- Leaders unsure how to measure success
- Technical teams overwhelmed by demand

This is not a technology problem. It's a strategy problem.

1.2 The Cost of Not Having a Strategy

When organizations skip strategy and jump straight into tools or pilots, predictable patterns emerge:

These failures are not inevitable. They are the result of skipping the foundational work of strategy and road mapping.

1.3 The Shift from Experiments to Adoption

For the past decade, AI lived in innovation labs, research teams, and isolated pockets of experimentation. That era is over. AI is now moving into:

- Core business enterprise operations
- Customer and citizen services
- Clinical and administrative workflows
- Program integrity and fraud detection
- Workforce productivity
- Decision support
- Contact centers
- Supply chain and logistics

This shift changes everything.

Experimentation tolerates ambiguity. Enterprise adoption demands clarity.

Managers are now responsible for:

- Prioritizing use cases
- Aligning AI to mission outcomes
- Managing risk
- Coordinating across teams
- Communicating change
- Ensuring responsible deployment
- Delivering measurable value

This is why AI strategy is no longer optional. It is the foundation for sustainable adoption.

1.4 Mgr's Evolving Role in an AI-Driven Org.

AI changes managers' roles in three fundamental ways.

1.5 From Project Owners to Portfolio Stewards

Managers must think beyond individual tools or pilots. They must manage a portfolio of AI opportunities, balancing:

Quick wins

- High-impact bets
- Risk mitigation
- Organizational readiness

2. From Technology Consumers to Workflow Designers

Managers don't need to understand model architecture. They do need to understand:

- How workflows change
- Where humans stay in the loop
- How decisions are made
- How data flows
- How risk is introduced and mitigated

AI is workflow technology. Managers own workflows.

3. From Change Observers to Change Leaders

AI adoption is a cultural transformation. Managers must lead teams through:

- Uncertainty
- Skill shifts
- New responsibilities
- New ways of working

This requires communication, empathy, and clarity—not technical expertise.

1.6 AI Without Strategy Creates Chaos

Let's be direct: AI is powerful enough to transform your organization and chaotic enough to break it.

Without a strategy, AI becomes:

- A distraction
- A cost center
- A risk multiplier
- A source of confusion
- A magnet for unrealistic expectations

With strategy, AI becomes:

- A force multiplier
- A driver of mission outcomes
- A catalyst for modernization
- A source of competitive advantage
- A foundation for responsible innovation

The difference is not technology. The difference is the leadership.

1.7 How AI Differs from Traditional IT Strategy

AI strategy is not an extension of the IT strategy. It requires a different mindset.

1. AI is probabilistic, not deterministic

Traditional IT systems follow rules. AI systems make predictions. This introduces uncertainty that must be managed.

2. AI changes workflows, not just tools

AI alters how work is done, not just which tools are used. This requires organizational and cultural change.

3. AI requires continuous iteration

AI systems improve over time. They require monitoring, evaluation, and refinement.

4. AI introduces new categories of risk

Bias, hallucinations, privacy, model drift—these are not traditional IT risks.

5. AI value depends on data quality

Data becomes a strategic asset rather than an afterthought.

6. AI requires cross-functional collaboration

1.8 AI Projects vs AI Transformation: The Difference

Many organizations confuse activity with progress. They run AI projects and assume they are transforming.

They are not.

AI Projects

- Isolated
- Tactical
- Tool-driven
- Short-term
- Often unscalable

AI Transformation

- Strategic
- Mission-aligned
- Workflow-driven
- Responsible by design
- Scalable and Sustainable

This book is about transformation, not projects.

1.9 Why begin the book with Strategy & Roadmapping?

- You cannot scale what you cannot explain.
- You cannot explain what you have not planned.
- You cannot plan what you have not aligned.

Strategy creates alignment. Roadmapping creates momentum; together, they create clarity.

This book begins with strategy and roadmapping because they are the foundation for everything that follows:

- Governance

- Use case selection
- Operating models
- Pilots
- Scaling
- Responsible AI
- Workforce transformation

Without a strategy, these become disconnected activities. With strategy, they become a coherent system.

1.10 Qualifications for Leading This Initiative

If you are reading this book, you are already positioned to lead AI adoption. You don't need to be a data scientist. You don't need to be a technologist. You need to be:

- Curious
- Mission-driven
- Outcome-focused
- Comfortable with ambiguity
- Willing to learn
- Willing to lead

AI strategy is not about technical expertise. It is about clarity, alignment, and leadership.

You can do this. This book will show you how.

1.11 Summary — Key Takeaways

- AI has moved from experimentation to enterprise adoption.
- Without a strategy, AI creates fragmentation, risk, and wasted effort.

- Managers—not technical teams—are now central to AI success.
- AI strategy is fundamentally different from traditional IT strategy.
- Roadmapping provides the structure and momentum needed for real adoption.
- AI transformation is about workflows, people, and mission—not just technology.
- You don't need to be a technical expert to lead AI strategy. You need clarity and a framework.

2 What AI Can (and Cannot) Do

Note that Chapter 2 is one of the most important foundations in the entire guide because it gives managers a clear, reality-based understanding of what AI is actually capable of today. The chapter divides AI into four core functions: classification, prediction, generation, and interaction, giving leaders clear insight into strengths, limits, and business impacts. This structure cuts through the noise of marketing claims and vendor promises, grounding your thinking in what these systems can reliably do, what they struggle with, and why. It also reframes AI not as magic, but as a set of tools with predictable behaviors, failure modes, and operational requirements.

Just as importantly, Chapter 2 equips you to separate what's real from hype and to understand the limits of current models—where they break, where they hallucinate, where they require human oversight, and where they introduce risk. This becomes the basis for evaluating feasibility without requiring a data scientist. Instead of relying on intuition or vendor persuasion, you learn to ask sharper questions, recognize technical red flags, and judge whether a use case is truly viable. For managers responsible for budgets, teams, and outcomes, this chapter builds the strategic literacy needed to make confident, defensible decisions about where AI can create value—and where it can't.

2.1 Opening Scenario: The Confusion Problem

- A director asks whether AI can summarize clinical notes.

- A program manager wants to automate eligibility decisions.
- A vendor claims their model can "understand any document."
- Your legal team warns about hallucinations.
- Your CIO says the data isn't ready.
- Your CEO wants to know why you're not moving faster.

Everyone is talking about AI. Very few are talking about what AI can *actually* do.

This chapter gives you the clarity you need to separate capability from hype, feasibility from fantasy, and opportunity from risk. Before you build a strategy, you must understand the technology's boundaries.

2.2 Why AI Managers Need a Clear Mental Model

You don't need to understand model architecture or training pipelines. You *do* need a practical, manager-ready way to evaluate:

- What AI is good at
- What AI struggles with
- Where AI introduces risk
- Where humans must stay in the loop
- How to judge feasibility without being a data scientist

This chapter gives you that mental model.

2.3 The Four Major Categories of AI Capabilities

AI is often described as a single technology, but managers should think of it as four distinct capability buckets. This helps you match the right tool to the right problem.

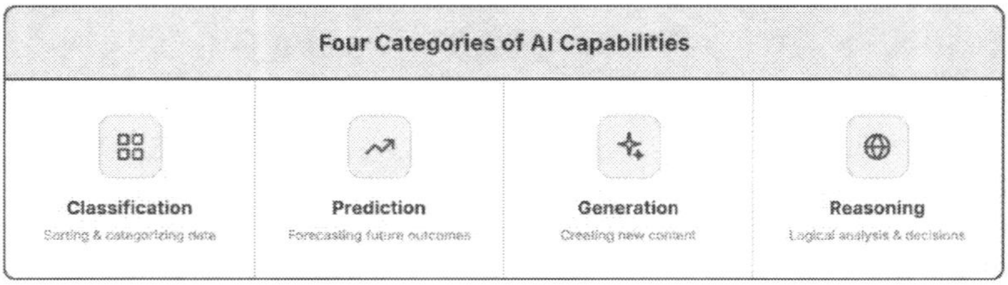

1. Classification: Sorting and Labeling Information

What it does: AI classifies, categorizes, or assigns labels to data.

Examples:

- Flagging fraudulent claims
- Categorizing documents
- Identifying sentiment in customer feedback
- Routing tickets to the right team

Strengths:

- Fast
- Consistent
- Scales easily

Limitations:

- Requires clear categories
- Can struggle with edge cases
- Needs high-quality training data

Manager's rule: If the task is "decide which bucket this goes into," AI is usually strong.

2. Prediction: Estimating What Will Happen Next

What it does: AI predicts outcomes based on patterns in historical data.

Examples:

- Predicting patient no-shows
- Forecasting call volume
- Estimating risk of equipment failure
- Predicting the likelihood of claim denial

Strengths:

- Excellent at pattern recognition
- Useful for planning and resource allocation

Limitations:

- Predictions are probabilistic, not certain
- Bias in historical data can distort results

Manager's rule: If the task is "what is likely to happen," AI can help—but humans must interpret.

3. Generation: Creating New Content

This is the category that exploded with the rise of large language models (LLMs).

What it does: AI generates text, summaries, drafts, images, or code.

Examples:

- Drafting emails or reports
- Summarizing long documents
- Creating patient instructions
- Generating code snippets
- Producing knowledge articles

Strengths:

- Fast content creation

- Reduces administrative burden
- Useful for first drafts

Limitations:

- Can hallucinate
- Needs guardrails
- Must be reviewed in high-stakes environments

Manager's rule: If the task is "create a draft," AI is powerful—but requires oversight.

4. Reasoning: Applying Logic to Solve Problems

This is the newest and most rapidly evolving capability.

What it does: AI chains steps, applies rules, and performs structured reasoning.

Examples:

- Multi-step workflows
- Policy-aware decision support
- Complex question answering
- Drafting responses based on rules or regulations

Strengths:

- Can handle more complex tasks than simple generation
- Useful for policy-driven environments

Limitations:

- Still maturing
- Requires careful prompt design and evaluation
- Not a replacement for human judgment

Manager's rule: If the task requires logic or rules, AI can assist—but not autonomously.

2.4 What AI Is Good At

Managers should achieve strengths that consistently hold across industries.

1. High-volume, repetitive tasks

AI excels at tasks humans find tedious:

- Summarizing
- Categorizing
- Drafting
- Extracting information
- Routing

These are ideal starting points.

2. Pattern recognition at scale

AI can detect patterns humans miss:

- Fraud signals
- Operational anomalies
- Behavioral trends
- Clinical patterns

This is where AI becomes a force multiplier.

3. First drafts, not final decisions

AI is exceptional at producing:

- Drafts
- Options
- Variations
- Summaries

Humans refined. AI accelerates.

4. Structured workflows with clear rules

AI thrives when:

- Inputs are consistent
- Rules are explicit
- Outcomes are well-defined

This is why document processing and customer service are early winners.

What AI Struggles With

Understanding limitations is just as important as understanding capabilities.

1. Ambiguous or poorly defined tasks

If humans can't agree on the rules, AI won't either.

2. Tasks requiring deep domain judgment

AI can support clinicians, analysts, and investigators—but not replace them.

3. Novel situations with no historical precedent

AI learns from patterns. If the pattern has never occurred, the AI cannot reliably infer.

4. Tasks with high risk and low tolerance for error

Examples:

- Medical diagnosis
- Legal decisions
- Eligibility determinations
- Safety-critical operations

AI can assist, but humans must own the decision.

5. Incomplete, inconsistent, or biased data

AI amplifies whatever data it is given. If the data is flawed, the output will be flawed.

The Limits of Current Models

Managers should internalize these boundaries:

AI does not understand content the way humans do.

It predicts the next best word or token.

AI can hallucinate.

It may produce confident but incorrect answers.

AI is not autonomous.

It cannot reliably make unsupervised decisions in high-stakes environments.

AI is sensitive to input phrasing.

Small changes in prompts can produce different results.

AI requires monitoring.

Models drift. Data changes. Workflows evolve.

These limitations are not defects—they are characteristics of technology.

2.5 So, you're not a Data Scientist

Managers can quickly assess feasibility using four questions:

1. Is the task repetitive or pattern-based?

If yes, AI is likely a good fit.

2. Are the inputs and outputs consistent?

If the workflow varies wildly, AI may struggle.

3. Is there enough data?

AI needs example documents, cases, transcripts, and logs.

4. What is the risk of being wrong?

High-risk tasks require human oversight.

The Manager's Rule of Thumb: AI Is a Tool, not a Teammate

AI is powerful, but it is not a replacement for:

- Judgment
- Empathy
- Accountability
- Ethical reasoning
- Mission ownership

AI augments humans. Humans remain responsible.

2.6 The Responsible AI Lens: Why Limits Matter

Understanding what AI cannot do is essential for responsible deployment. It helps you:

- Set realistic expectations
- Avoid over-automation
- Protecting vulnerable populations
- Maintain trust with employees and customers
- Reduce legal and compliance risk

Responsible AI begins with clarity about capability boundaries.

2.7 Summary — Key Takeaways

- AI capabilities fall into four categories: classification, prediction, generation, and reasoning.
- AI is strong at repetitive tasks, pattern recognition, and first-draft creation.
- AI struggles with ambiguity, high-risk decisions, and tasks requiring deep domain judgment.
- AI does not "understand" content; it predicts patterns.
- Managers can evaluate feasibility using four simple questions.
- AI is a workflow, not a replacement for human judgment.
- Understanding limitations is essential for responsible, mission-aligned AI adoption.

2.8 Companion Workbook: Chapters 1 and 2

Section Overview

The Chapters 1 and 2 establish the foundation for your AI strategy work:

- **Ch. 1**: Why AI strategy matters and the leadership role you play
- **Ch. 2**: What AI can and cannot do — a practical capability model

This workbook section helps you internalize those concepts and apply them to your organization immediately.

Workbook Section for Chapter 2

The AI Moment: Understanding Your Organization's Starting Point

Exercise 1: The AI Pressure Map

Every organization feels pressure to "do something with AI," but the sources of that pressure differ. Use this map to identify the sources of pressure in your environment.

Instructions:

List the top 3–5 sources of AI pressure in your organization.

Table 1 Pressure Evaluation Table

Source of Pressure	What They're Asking For	Why it Matters	Level of Urgency (Low/Medium/High)

Reflection:

Which pressures are legitimate drivers of strategy, and which are noise?

Exercise 2: AI Fragmentation Assessment

Use this quick diagnostic to identify early signs of fragmentation.

Check all that apply:

- ☐ Multiple teams experimenting independently
- ☐ No shared definition of "AI success."
- ☐ Vendors pitching conflicting solutions
- ☐ Pilots that never scaled
- ☐ Shadow AI (unapproved tools)
- ☐ No clear owner for AI governance
- ☐ Leaders asking for results without clarity on goals

If you checked 3 or more: Your organization is at high risk of AI fragmentation — strategy is urgent.

Exercise 3: Your Role in AI Leadership

Managers are now central to AI adoption. Clarify your role using the prompts below.

Prompts:

- What decisions about AI fall within your responsibility?
- What decisions do you influence but not own?
- What decisions are unclear or contested?

Write your responses:

My responsibilities:

____

My areas of influence:

____

Unclear or contested areas:

____

Reflection:

Where do you need clearer authority or alignment?

____

Exercise 4: AI Strategy Readiness Snapshot

Rate your organization on the following dimensions (1 = low, 5 = high).

Dimension	Score (1-5)	Notes
Leadership Alignment		
Data Readiness		
Workforce Readiness		
Process maturity		
Culture & change mgmt.		
Governance & Risk Mgmt.		

Figure 1 Readiness Assessment

Reflection:

Which two areas require the most attention before scaling AI?

Exercise 5: AI Capability Mapping

Use the four AI capability categories to map potential opportunities.

Instructions:

List 2–3 tasks in your organization that fit each category.

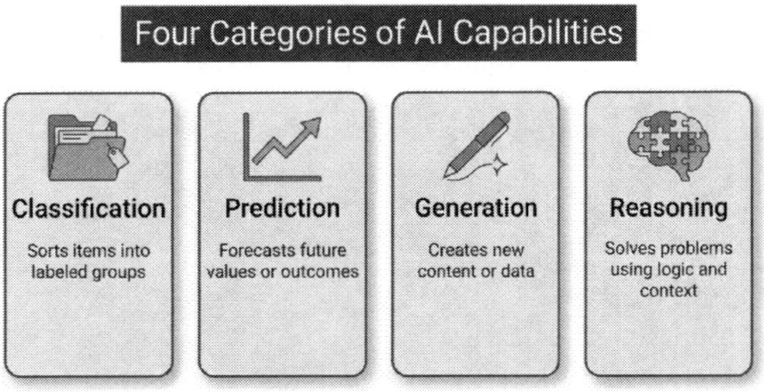

Table 1 Classification Projects

Task	Notes on Feasibility
1.	
2.	
3.	

Table 2 Prediction Projects

Task	Notes on Feasibility
1.	
2.	
3.	

Table 3 Generation Projects

Task	Notes on Feasibility
1.	
2.	
3.	

Table 4 Reasoning Projects

Task	Notes on Feasibility
1.	
2.	
3.	

Reflection:

Which category appears most promising for early wins?

Exercise 6: The AI Fit Test

Choose one workflow or task you're considering for AI. Evaluate it using the four feasibility questions.

Task/Workflow:

____

Feasibility Questions:

1.**Is the task repetitive or pattern-based?**

Notes:

____

2.**Are the inputs and outputs consistent?**

Notes:

____

3. **Is there enough data to learn from?**

Notes:

____

4. **What is the risk of being wrong?**

Notes:

____

Conclusion:

Is this a good candidate for AI?

- ☐ Yes
- ☐ Maybe (needs guardrails)
- ☐ No

Exercise 7: AI Limitations Risk Scan

Understanding limitations helps you avoid over-automation.

Instructions:

For each limitation, identify one workflow where this risk is relevant.

AI Limitation	Workflow Impacted	Risk Description
Ambiguous		
High-Stakes Decisions		
Novel Situations		
Poor Data Quality		
Need for Deep Domain Judgment		

Reflection:

Which risks require a human-in-the-loop design?

Exercise 8: Human + AI Workflow Design

AI augments humans — it doesn't replace them. Design a simple Human + AI workflow for a task in your organization.

Template:

Task:

____

AI does:

- ____
- ____
- ____

Humans do:

- ____
- ____

•

Why does this division work:

Workbook Summary: Chapters 2

By completing these exercises, you should now have:

- A clear picture of the pressures and expectations around AI in your organization
- An early assessment of fragmentation risks
- A defined sense of your leadership role in AI adoption
- A practical understanding of what AI can and cannot do
- A shortlist of feasible AI opportunities
- A clearer view of where humans must remain central

These insights will feed directly into the next workbook sections on:

- Use case prioritization
- Readiness assessment
- Operating model design
- Building your 90-day roadmap

3 The AI Value Equation

3.1 Scenario: The "Show Me the Value" Moment

You've been asked to brief your leadership team on AI opportunities. You walk into the room prepared to talk about capabilities, risks, and use cases. But the first question you get is the hardest one:

"What's the actual value?"

- Not a theoretical value.
- Not potential value.
- Not the vendor's-promised value.
- Real, measurable, operational value.

This is the moment where many AI initiatives stall. Not because the technology isn't ready — but because leaders can't clearly articulate the value equation.

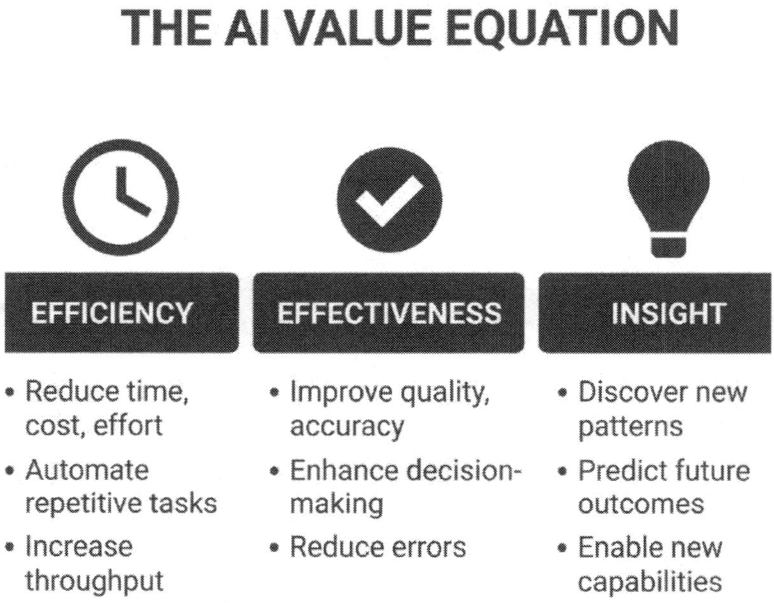

This chapter gives you the framework to do exactly that.

3.2 Why Value Must Come Before Use Cases

Most organizations start with use cases:

- "We should automate this workflow."
- "We should summarize these documents."
- "We should build a chatbot."

But use cases without a value framework lead to:

- Misaligned priorities
- Low-impact pilots
- Overinvestment in the wrong areas

- Underinvestment in the right ones
- Difficulty securing executive support

Value must come first.

Use cases come second.

A clear value framework helps you:

- Prioritize what matters
- Say no to low-value ideas
- Communicating impact to leadership
- Build a roadmap that drives outcomes
- Avoid chasing hype

3.3 The Three Drivers of AI Value

AI creates value in only three ways.

Everything else is noise.

edo in blue, make it modern, add imagery

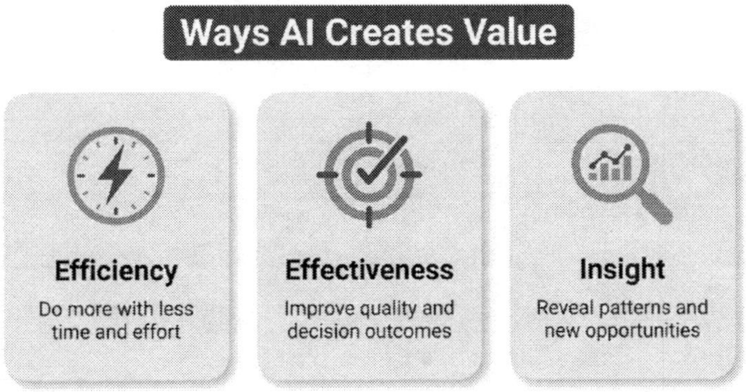

1. Efficiency: Doing the Same Work Faster or Cheaper

This is the most common and easiest to measure.

Examples:

- Reducing administrative burden
- Automating repetitive tasks
- Accelerating document processing
- Reducing time spent on drafting or summarizing
- Streamlining customer or patient interactions

The efficiency value is operational. It shows up in:

- Hours saved
- Cost avoided
- Cycle time reduced
- Throughput increased

This is where most organizations start — and where early wins live.

2. Effectiveness: Doing the Work Better

This is where AI begins to enhance quality, accuracy, and decision-making.

Examples:

- Improving triage accuracy
- Enhancing fraud detection
- Reducing errors in manual processes
- Supporting clinicians or analysts with better insights
- Improving customer or citizen experience

Effectiveness value is quality-driven.

It shows up in:

- Fewer errors

- Better decisions
- Higher satisfaction
- Improved outcomes

This is where AI becomes a force multiplier.

3. Insight: Doing Work You Couldn't Do Before

This is the most transformative — and the hardest to quantify.

Examples:

- Identifying patterns humans can't see
- Predicting risk before it materializes
- Surfacing insights from millions of documents
- Enabling new mission capabilities
- Supporting strategic planning with data-driven foresight

Insight value is strategic. It shows up in:

- New capabilities
- Better resource allocation
- Earlier interventions
- Improved mission outcomes

This is where AI becomes a competitive advantage.

3.4 The AI Leverage Curve

Not all tasks benefit equally from AI. Some tasks see marginal improvement. Others see exponential improvement.

The AI Leverage Curve helps you identify where AI creates disproportionate value.

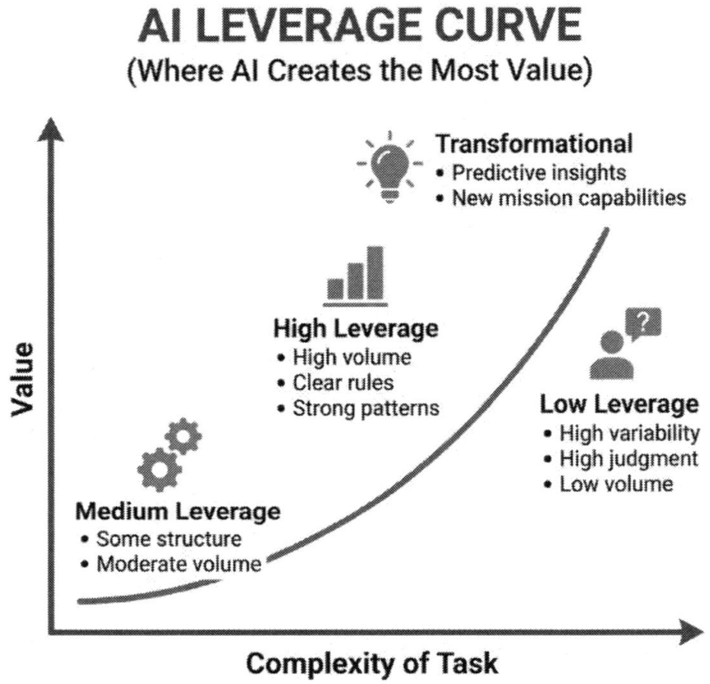

AI can unlock entirely new capabilities.

Your job as a manager is to identify where your workflows sit on this curve.

3.5 How to Identify High-Value AI Opportunities

Use this simple three-part test.

1. Volume: How often does the task occur?

High-volume tasks produce high returns.

Examples:

- Claims
- Tickets

- Messages
- Documents
- Encounters
- Calls

If a task happens thousands of times per month, AI can create meaningful value.

2. Variability: How consistent is the task?

Low variability = high AI potential. High variability = human-centric.

Ask:

- Are the inputs human-centric? Are the rules clear?
- Are the outputs predictable?

If the answer is yes, AI can help.

3. Risk: What happens if the AI is wrong?

Risk determines the level of human oversight required.

- **Low risk:** automate
- **Medium risk:** human-in-the-loop
- **High risk:** AI assists, humans decide

Risk is not a blocker — it's a design constraint.

CCase Examples Across Industries

Healthcare

- **Efficiency:** Automating prior authorization documentation
- **Effectiveness:** Improving clinical note summarization
- **Insight:** Predicting patient deterioration

Public Sector

- **Efficiency:** Automating citizen inquiry responses
- **Effectiveness:** Enhancing program integrity reviews
- **Insight:** Identifying emerging fraud patterns

Enterprise

- **Efficiency:** Drafting customer communications
- **Effectiveness:** Improving supply chain forecasting
- **Insight:** Detecting operational anomalies

These examples show how the same value framework applies across domains.

3.6 The Manager's Value Checklist

Use this checklist to evaluate any proposed AI idea.

Value

- ☐ Does this reduce time, cost, or effort?
- ☐ Does this improve quality or accuracy?
- ☐ Does this unlock new insights or capabilities?

Feasibility

- ☐ Is the task repetitive or pattern-based?
- ☐ Are inputs and outputs consistent?
- ☐ Is there enough data?

Risk

- ☐ What happens if the AI is wrong?
- ☐ What oversight is required?
- ☐ What safeguards are needed?

Impact

- ☐ How many people or workflows benefit?
- ☐ How often does the task occur?
- ☐ How measurable is the outcome?

If you can't check at least one value box and one feasibility box, the idea is not ready.

3.7 The Responsible AI Lens: Value Without Harm

Value is not just about impact — it's about responsible impact.

Ask:

- Who benefits?
- Who might be harmed?
- What biases could be amplified?
- What safeguards are needed?
- How transparent is the workflow?

Responsible AI is not a constraint. It is a requirement for sustainable value.

3.8 Summary — Key Takeaways

- AI creates value in only three ways: efficiency, effectiveness, and insight.
- High-value opportunities sit at the intersection of volume, consistency, and manageable risk.
- The AI Leverage Curve helps you identify where AI can create disproportionate impact.
- Value must come before use cases — it is the foundation of prioritization.
- Managers must evaluate value, feasibility, risk, and impact before committing to an AI initiative.

- Responsible AI ensures value is delivered safely and sustainably.

3.9 Companion Workbook 3: The AI Value Equation

This workbook section helps you apply the concepts from Chapter 3 to your own organization, enabling you to identify high-value AI opportunities with clarity and confidence.

Section Overview

Ch 3 introduced:

- The **three drivers of AI value**
- The **AI Leverage Curve**
- How to identify **high-value opportunities**
- How to evaluate **volume, variability, and risk**
- The **Manager's Value Checklist**

This workbook turns those ideas into actionable exercises.

Exercise 1 — Map Your Current AI Value Opportunities

Use the table below to categorize potential AI opportunities by the three value drivers.

Instructions:

List 3–5 workflows or tasks your organization performs today.

Then identify which type of value each one could generate.

Workflow/Task	Efficiency Value	Effectiveness Value	Insight Value	Notes
	☐	☐	☐	
	☐	☐	☐	

	☐	☐	☐	
	☐	☐	☐	

Reflection:

Which workflows show the strongest value potential?

Exercise 2 — The AI Leverage Curve Assessment

Place each workflow on the AI Leverage Curve.

Instructions:

For each workflow, check the category that best fits.

Workflow/Task	Low Leverage	Medium Leverage	High Leverage	Transformational
	☐	☐	☐	☐
	☐	☐	☐	☐
	☐	☐	☐	☐
	☐	☐	☐	☐

Reflection:

Which workflows fall into the **High** or **Transformational** categories?

These are your strongest candidates for early AI investment.

Exercise 3 — Volume, Variability, and Risk Scoring

Use this scoring model to evaluate feasibility and impact.

Instructions:

Score each workflow on each dimension from 1 to 5.

Workflow/Task	Low Leverage	Medium Leverage	High Leverage	Transformational

Risk is reverse-scored:

- **1 = low risk** (ideal for automation)
- **5 = high risk** (AI should assist, not decide)

Reflection:

Which workflows have:

- **High volume**
- **Low variability**
- **Low to moderate risk**

These are your best near-term opportunities.

Exercise 4 — The Manager's Value Checklist (Applied)

Choose one workflow and evaluate it using the checklist.

Selected Workflow:

Value

- ☐ Reduces time, cost, or effort
- ☐ Improves quality or accuracy
- ☐ Unlocks new insights or capabilities

Feasibility

- ☐ Repetitive or pattern-based
- ☐ Consistent inputs and outputs

- ☐ Sufficient data available

Risk

- ☐ Acceptable risk if AI is wrong
- ☐ Clear oversight model
- ☐ Safeguards can be implemented

Impact

- ☐ Benefits many people or workflows
- ☐ Occurs frequently
- ☐ Measurable outcomes

Conclusion:

Is this workflow a strong candidate for AI?

- ☐ Yes
- ☐ Maybe (needs guardrails)
- ☐ No

Exercise 5 — Draft Your AI Value Narrative

This is the narrative you'll eventually use with leadership.

Prompts:

- What value will AI create?
- Who benefits?
- How will you measure success?
- What risks must be managed?
- Why now?

Write your narrative:

____

Workbook Summary — Chapter 3

By completing this section, you now have:

- A clear map of where AI can create efficiency, effectiveness, and insight
- A leverage-based view of your workflows
- A feasibility and risk score for each opportunity
- A shortlist of high-value AI candidates
- A draft narrative to communicate value to leadership

Part II – Building an AI Strategy that Works

Part II, *Building an AI Strategy That Works*, moves leaders from understanding the technology to shaping a plan that actually delivers results. This section focuses on the practical disciplines that separate successful AI initiatives from those that stall—clear mission alignment, rigorous problem definition, workflow-level analysis, and a structured approach to prioritizing use cases. Instead of starting with tools, models, or vendor offerings, Part II teaches managers how to anchor every AI decision in the organization's goals and constraints. It shows how to translate strategic objectives into AI-enabled outcomes, how to identify the operational problems that matter most, and how to ensure AI is being applied where it can create measurable value.

Just as importantly, Part II helps leaders avoid the most common failure mode in AI adoption: technology-first thinking. Through frameworks like the "Mission → Problem → Workflow → AI" chain, this section builds the discipline of starting with purpose, not novelty. It guides managers through evaluating feasibility, sequencing initiatives, and designing an AI portfolio that is realistic, aligned, and resilient. By the end of Part II, leaders have a strategy they can defend, communicate, and execute—one that fits their mission, data, workflows, and people, rather than chasing trends or reacting to pressure.

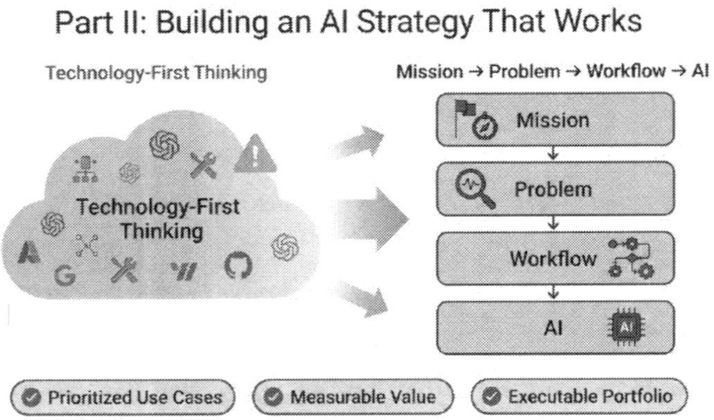
Part II: Building an AI Strategy That Works
Technology-First Thinking
Mission → Problem → Workflow → AI
Technology-First Thinking
Mission
Problem
Workflow
AI
Prioritized Use Cases
Measurable Value
Executable Portfolio

4 Start with Mission, Not Models

It is important to note that Chapter 4 marks a pivotal shift in this guide, redirecting focus from pursuing technological advancements for their own sake to grounding AI strategy in the organization's mission. Aligning AI initiatives with the organizational mission has consistently proven to be the most reliable indicator of success, surpassing factors such as model scale, algorithmic complexity, or the number of pilot projects. By establishing mission alignment at the outset, leaders ensure that AI addresses meaningful challenges—optimizing outcomes, enhancing operations, and supporting strategic objectives. This chapter provides guidance on translating broad organizational goals into concrete, measurable AI results, offering managers an effective framework for bridging ambition and execution.

The chapter also introduces the "Mission → Problem → Workflow → AI" chain, a simple but powerful discipline that prevents teams from falling into tech-first thinking. By tracing every AI idea back to a real mission need, a real operational problem, and a real workflow where that problem occurs, leaders avoid wasted investments and unfocused experimentation. This approach helps managers distinguish between attractive technologies and meaningful solutions, ensuring AI is deployed where it can create durable value.

4.1 Opening Scenario: The Wrong Starting Point

A senior leader announces that the organization needs to "get serious about AI."

Within days:

- Vendors begin pitching solutions
- Teams start experimenting with tools
- Executives ask for quick wins
- Analysts build prototypes
- IT worries about risk and governance

Everyone is moving fast.

No one is moving together.

This is the most common failure pattern in AI adoption:

"Starting with technology instead of Mission."

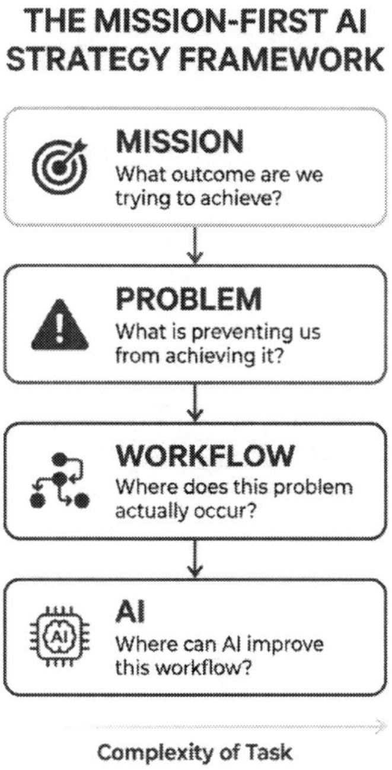

AI strategy does not begin with models, tools, or vendors. It begins with clarity about what your organization exists to do — and how AI can help you do it better.

This chapter gives you the framework to anchor AI in mission, not hype.

Why Mission Alignment Is the #1 Predictor of AI Success

Organizations that succeed with AI share one trait:

They tie every AI initiative to a mission-critical outcome.

Organizations that fail share the opposite trait:

They chase technology without a clear purpose.

Mission alignment matters because it:

- Focuses teams on outcomes, not features
- Prevents wasted investment
- Reduces risk by grounding decisions in real needs
- Accelerates adoption because people understand the "why."
- Creates executive clarity and support
- Ensures AI enhances — not distracts from — core responsibilities

AI is not a strategy.
AI is a tool that supports strategy.

Table 5 The Right Order for an AI tool supporting Corporate Strategy

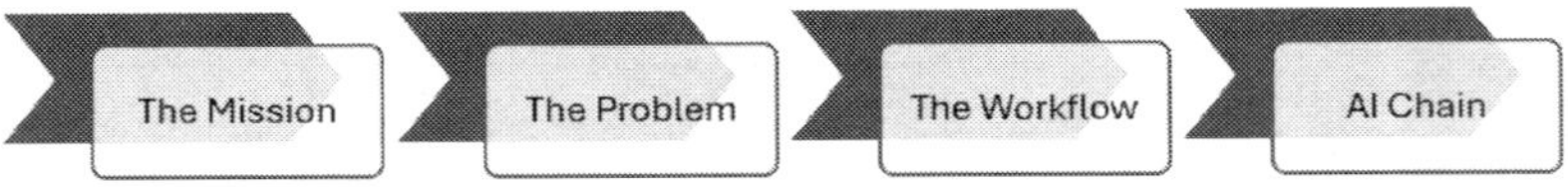

The Mission → Problem → Workflow → AI Chain

This is one of the most important frameworks in the entire book. It ensures AI is applied where it matters — and only where it matters.

Most organizations start at the end:

Table 6 Wrong order for AI

("We have a model — where can we use it?")

4.2 Successful organizations start at the beginning:

Let's break it down.

1. Mission: What outcome are we trying to achieve?

This is the anchor for everything.

Examples:

- Improve patient outcomes
- Reduce fraud, waste, and abuse
- Enhance readiness
- Improve citizen experience
- Reduce administrative burden
- Increase operational efficiency

If you cannot articulate the mission outcome, you are not ready for AI.

2. Problem: What is preventing us from achieving that mission?

This is where clarity begins.

Examples:

- Clinicians spend too much time on documentation
- Investigators cannot review all cases
- Citizens wait too long for responses
- Analysts are overwhelmed by data
- Processes are inconsistent across teams

AI should never solve a problem that does not exist.

3. Workflow: Where does the problem actually occur?

AI is workflow technology. If you cannot point to the workflow, you cannot apply AI effectively.

Examples:

- Intake and triage

- Document review
- Case investigation
- Customer service
- Claims processing
- Scheduling
- Reporting

Workflows are where value is created — or lost.

4. AI: What capability can improve the workflow?

Only now do we talk about AI.

Examples:

- Classification to route cases
- Generation to draft summaries
- Prediction to identify risk
- Reasons to apply the policy
- Automation to reduce manual steps

AI is the last step — not the first.

Table 7 The Mission-First AI Strategy Framework

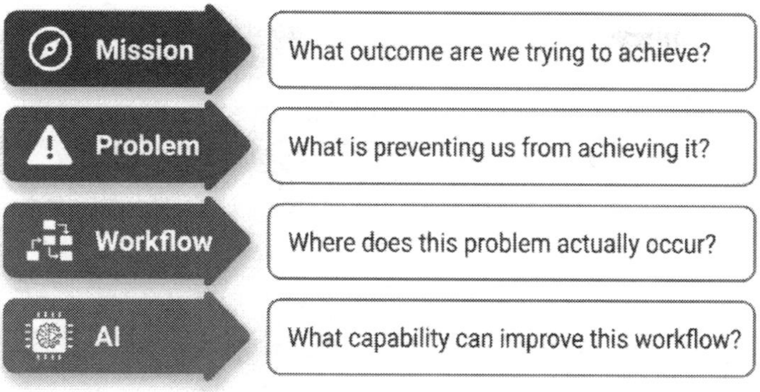

4.3 Why Starting with AI Leads to Failure?

When organizations start with technology, predictable failure patterns emerge:

Table 8 Mistake of Starting with AI

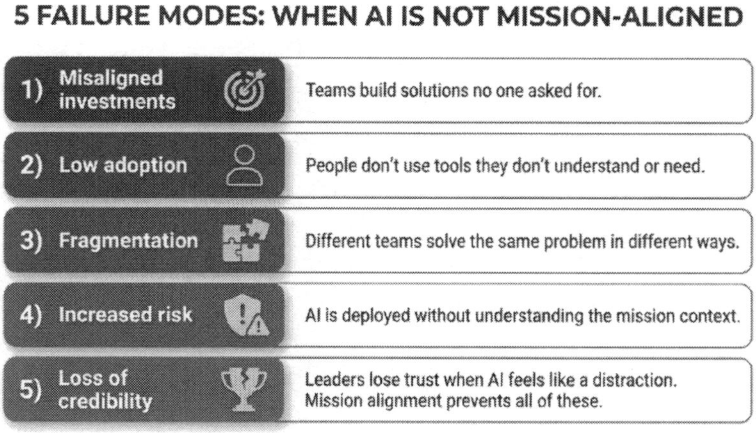

4.4 How to Translate Mission Into AI Opportunities

Use this three-step process.

Step 1: Identify Mission-Critical Outcomes

Ask:

- What outcomes matter most this year?
- What outcomes are underperforming?
- What outcomes have the highest strategic importance?

Examples:

- Reduce clinician burnout
- Improve citizen satisfaction
- Increase readiness
- Reduce backlog
- Improve program integrity

Step 2: Identify the Problems Blocking Those Outcomes

Ask:

- What slows us down?
- What creates errors?
- What frustrates staff or customers?
- What consumes the most time?
- What creates the most risk?

Step 3: Identify the Workflows Behind Those Problems

Ask:

- Where does the work actually happen?
- Who performs it?
- What steps are involved?
- What data is used?
- Where are the bottlenecks?

Only after these steps do you ask:

"Can AI help?"

4.5 Case Example Full Walk-Through: Healthcare

Mission: Improve patient outcomes

Improving patient outcomes is the foundational mission that guides nearly every clinical and operational decision in healthcare. It reflects the organization's commitment to delivering safer, more effective, and more equitable care while reducing preventable harm and unnecessary variation. At this level, leaders must address key factors such as timely care, accurate diagnosis, coordination, and clinician capacity, rather than focusing solely on technology. When framed correctly, the mission becomes a north star that aligns stakeholders, clarifies priorities, and ensures that any AI initiative is evaluated not by novelty or technical sophistication, but by its measurable contribution to better health, faster recovery, and improved patient experience.

Problem: Clinicians spend too much time on documentation

Excessive documentation is a major issue in healthcare, consuming hours of clinician time daily and leading to burnout, cognitive overload, and less time with patients. The problem is not simply that documentation takes time—it's that the time required often displaces the activities that most directly influence outcomes, such as patient communication, clinical reasoning, and care planning. This burden also creates downstream inefficiencies: delayed notes, incomplete information, inconsistent documentation quality, and increased administrative overhead. By articulating the problem in operational terms—lost clinical capacity, workflow friction, and quality variability—leaders can more clearly see where AI can relieve pressure without compromising safety or compliance.

Workflow: Clinical note creation and summarization

Creating and summarizing clinical notes is central to documentation and impacts patient care, billing, compliance, and clinical decisions. The workflow is structured and mentally demanding: clinicians combine patient history, exam results, diagnostic reasoning, and care plans into clear, standardized notes. Since much of this work is repetitive and time-sensitive, it lends itself well to augmentation. Breaking down each workflow step—gathering patient input, checking notes, adding new content, reconciling data, and finalizing records—shows where AI can simplify tasks, speed up processes, and standardize results. Understanding the workflow at this level of granularity ensures that AI is deployed where it can meaningfully enhance clinical efficiency without disrupting established clinical practices.

AI Capability: Generation + summarization

Generation and summarization capabilities enable AI systems to transform raw clinical inputs—speech, text, structured data—into coherent, accurate, and context-aware documentation. These models can draft encounter notes, summarize long histories, extract key findings, and standardize language to meet clinical and regulatory expectations. When implemented responsibly, they act as cognitive extenders, reducing manual effort while preserving clinician oversight and judgment. The value of these capabilities lies not only in speed but in consistency: AI can help ensure that critical details are captured, that documentation aligns with organizational standards, and that clinicians spend more time validating and refining content rather than producing it from scratch. By pairing generative capabilities with strong guardrails—human review, auditability, and domain-specific tuning—organizations can safely unlock efficiency gains while maintaining trust and clinical integrity.

What This Case Study Demonstrates

- ✓ Clear linkage between mission, operational pain point, workflow friction, and AI capability.
- ✓ Shows how generative AI augments—not replaces—clinical judgment.
- ✓ Illustrates a repeatable pattern leaders can apply across departments.

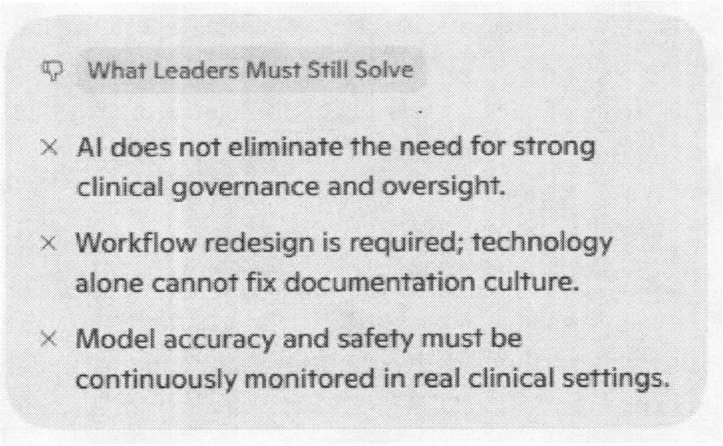

4.6 One-Page Handout: Clinical Documentation Augmentation with AI

Purpose

Equip managers and clinical leaders with a concise overview of how generative AI can reduce documentation burden while improving patient outcomes.

At a Glance

- **Mission:** Improve patient outcomes
- **Primary Problem:** Excessive clinician time spent on documentation
- **Target Workflow:** Clinical note creation and summarization
- **AI Capability:** Generation + summarization

Why This Matters

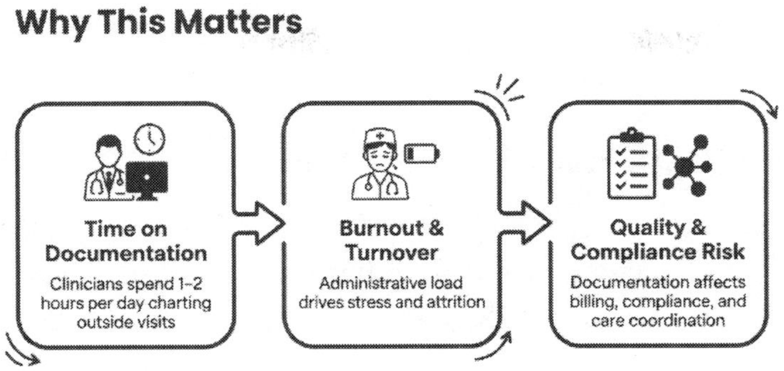

Where AI Helps

- Drafting encounter notes from transcripts
- Summarizing long patient histories
- Extracting key findings from structured and unstructured data
- Standardizing documentation language

Guard Rail Required

- Human-in-the-loop review
- Clinical governance and auditability
- Bias and hallucination monitoring
- Clear escalation paths for errors

Expected Outcomes

- 30–50% reduction in documentation time
- More face-to-face patient interaction
- Higher note consistency and completeness
- Improved clinician satisfaction

4.7 Case Study Full Walk-Through: Public Sector

Improving Citizen Experience with AI-Enabled Inquiry Routing

4.8 Mission: Improve Citizen Experience

Improving citizen experience is a core mission for modern public sector organizations, reflecting the expectation that government services should be accessible, responsive, and equitable. Citizens increasingly compare public services to the speed and convenience of private-sector digital experiences, creating pressure on agencies to reduce friction and deliver timely, accurate information. A mission framed around citizen experience shifts the focus from internal efficiency alone to the broader impact on trust, transparency, and public satisfaction. It encourages leaders to prioritize interventions that reduce delays, simplify interactions, and ensure that every citizen—regardless of channel or background—receives clear, consistent, and respectful service.

4.8.1 Problem: Long Wait Times for Responses

Long wait times for responses are one of the most visible and frustrating pain points for citizens engaging with government agencies. High inquiry volume, limited staffing, and complex routing rules often lead to backlogs that delay access to essential services, such as benefits, permits, and case updates. These delays erode public trust, increase repeat inquiries, and create additional administrative burden as citizens seek alternative channels to get answers. By articulating the problem in operational terms—queue congestion, inconsistent triage, and slow routing—leaders can see how the bottleneck is not simply a staffing issue but a workflow challenge that can be meaningfully improved with AI-enabled automation.

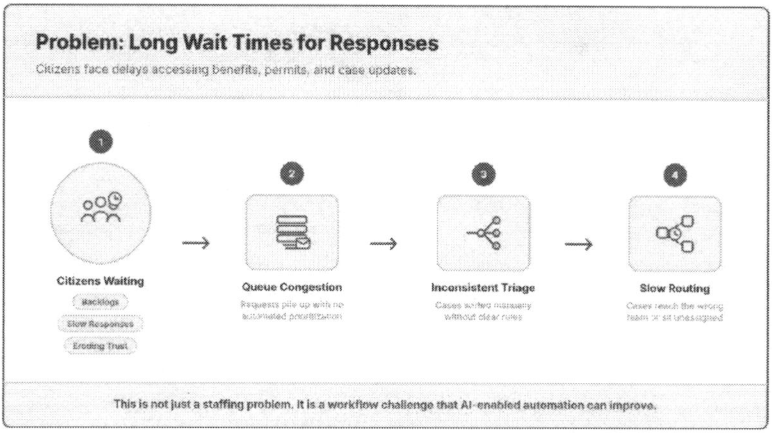

Figure 2 Long Wait Times

4.9 Workflow: Inquiry Intake and Routing

Inquiry intake and routing is a multi-step workflow that begins when a citizen submits a question through a portal, email, phone transcript, or in-person interaction. Staff must interpret the request, classify its intent, verify eligibility or context, and route it to the correct department or specialist. This process is often manual, repetitive, and highly variable, depending on staff experience and workload. Mapping the workflow reveals friction points, including inconsistent categorization, redundant data entry, and delays caused by unclear or incomplete citizen submissions. Because the workflow is structured, high-volume, and rules-driven, it is an ideal candidate for AI augmentation that accelerates triage while preserving human oversight for complex or sensitive cases.

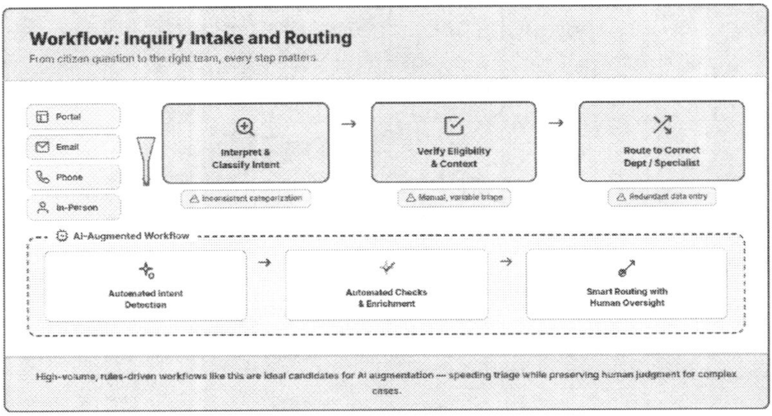

Figure 3 Intake and Routing

4.10 AI Capability: Classification + Generation

Classification and generation capabilities allow AI systems to automatically interpret citizen inquiries, categorize them into predefined service types, and draft initial responses or routing summaries. Classification models can detect intent, urgency, and required documentation, while generative models can produce clear, consistent, and policy-aligned messages that staff can review and approve. These capabilities reduce manual triage time, improve routing accuracy, and ensure that citizens receive timely, high-quality communication. When paired with strong governance—such as human review, audit trails, and policy-aligned templates—AI becomes a force multiplier that enhances service delivery without compromising fairness, accuracy, or accountability.

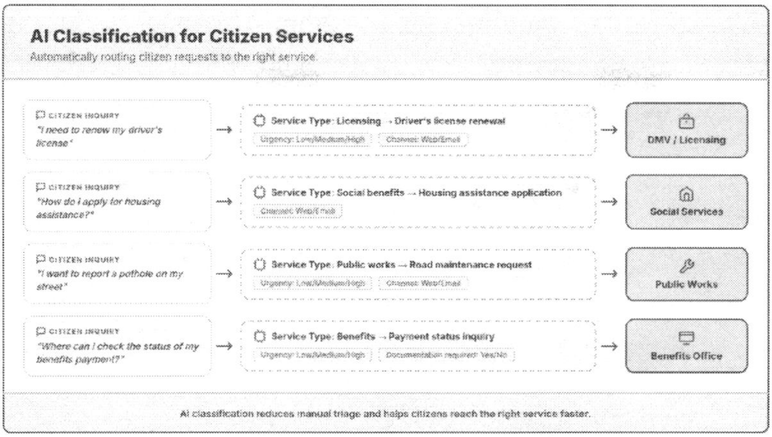

Full Case-Study Page

4.11 Public Sector Case Study: AI-Enabled Inquiry Intake and Routing

Context - Government agencies face rising citizen expectations and increasing inquiry volume across digital and traditional channels. Long response times undermine trust and create operational strain. This case study demonstrates how the Mission → Problem → Workflow → AI Capability framework guides leaders toward targeted, responsible AI adoption.

Mission - Improve citizen experience by delivering faster, clearer, and more consistent responses across all service channels.

Problem: Long wait times for responses due to manual triage, inconsistent routing, and high inquiry volume.

Workflow - Inquiry intake and routing, including classification, triage, verification, and assignment to the correct department or specialist.

AI Capability - Classification + generation to automatically categorize inquiries and draft initial responses or routing summaries.

4.12 One-Page Handout: AI for Citizen Inquiry Intake & Routing

Overview - A concise reference for public sector leaders evaluating AI opportunities in citizen-facing service workflows.

Mission: Improve the citizen experience through faster, more consistent, and more transparent communication.

Primary Problem - Long wait times for responses due to manual triage and routing bottlenecks.

Target Workflow - Inquiry intake and routing across email, portals, phone transcripts, and in-person submissions.

AI Capabilities

- **Classification:** Detect intent, urgency, and required service category.
- **Generation:** Draft initial responses, summaries, and routing notes for staff review.

Benefits

- Faster response times
- Reduced backlog and repeat inquiries
- More consistent communication
- Improved staff efficiency and morale

- Increased citizen trust and satisfaction

Required Guardrails

- Human-in-the-loop approval
- Policy-aligned templates
- Auditability and traceability
- Bias monitoring in classification
- Clear escalation paths for sensitive cases

Expected Outcomes

- 25–60% reduction in triage time
- Higher routing accuracy
- More predictable service levels
- Improved citizen experience metrics

Case Study: Enterprise — Reducing Claims Review Costs with AI-Driven Document Intelligence

Mission: Reduce Operational Costs

Reducing operational costs is a central mission for enterprise organizations seeking to remain competitive while improving service quality. Cost reduction is not simply about cutting expenses—it is about eliminating inefficiencies, reducing rework, and reallocating human expertise to higher-value activities. When framed correctly, this mission encourages leaders to identify processes where manual effort is high, variability is costly, and automation can create a measurable financial impact. It also ensures that AI investments are evaluated not by novelty but by their ability to streamline operations, reduce cycle times, and improve the accuracy and consistency of core business processes.

Problem: High Manual Effort in Claims Review

Claims review is one of the most labor-intensive processes in many enterprise environments, particularly in insurance, financial services, and warranty operations. Analysts must sift through large volumes of documents—forms, receipts, reports, correspondence—to determine eligibility, completeness, and next steps. This manual effort increases operational costs, slows processing times, and introduces variability in decision quality. The problem is not only the volume of claims but the complexity and inconsistency of the documents themselves. By articulating the problem as excessive manual review effort, leaders can clearly see how document intelligence and automation can reduce cost drivers while improving throughput and accuracy.

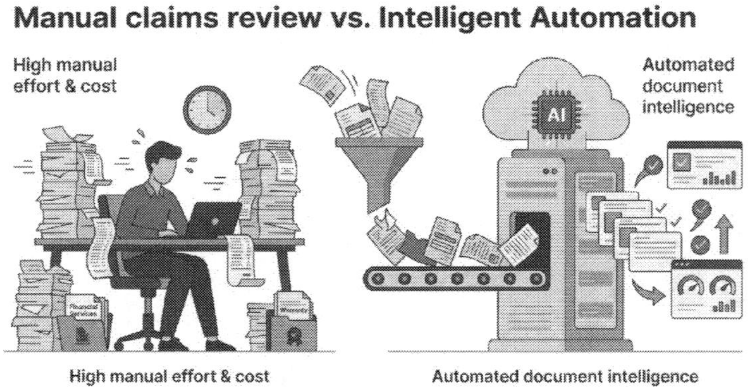

Workflow: Document Analysis and Triage

Document analysis and triage is a structured yet highly repetitive workflow that begins when a claim is submitted and ends when it is routed to the appropriate team or decision path. Analysts must extract key data points, validate supporting evidence, classify claim types, and identify missing or inconsistent information. This workflow is ideal for AI augmentation because it involves predictable patterns, high document volume, and well-defined decision rules. Mapping the workflow reveals friction points, including manual data extraction, inconsistent classification, and delays due to incomplete submissions. These steps consume significant analyst time and create bottlenecks that directly increase operational costs.

AI Capability: Classification + Extraction

Classification and extraction capabilities allow AI systems to automatically categorize claims, identify claim types, and extract key data fields from unstructured documents. These models can process large volumes of PDFs, images, and forms, turning them into structured data that accelerates triage and reduces manual review. Classification models determine the claim category or required workflow, while extraction models pull out essential details such as dates, amounts, identifiers, and supporting evidence. When paired with human oversight, auditability, and domain-specific tuning, these capabilities significantly reduce manual effort, improve accuracy, and shorten processing times—directly supporting the mission of reducing operational costs.

Full Case-Study Page

Enterprise Case Study: AI-Enabled Claims Review Automation

Context

Enterprises face rising operational costs driven by high document volume, complex claims, and manual review processes. Claims review is a prime candidate for AI augmentation because it is repetitive, rules-driven, and dependent on accurate document interpretation. This case study demonstrates how the Mission → Problem → Workflow → AI Capability framework guides leaders toward targeted, responsible AI adoption.

Mission

Reduce operational costs by streamlining claims review and reallocating analyst time to higher-value activities.

Problem

High manual effort in claims review due to document complexity, inconsistent formats, and repetitive data extraction.

Workflow

Document analysis and triage, including classification, data extraction, validation, and routing.

AI Capability

Classification + extraction to automatically categorize claims and extract key data fields from unstructured documents.

One-Page Handout: AI for Claims Review Automation

Overview

A concise reference for enterprise leaders evaluating AI opportunities in claims processing and document-heavy workflows.

Mission

Reduce operational costs by accelerating claims review and minimizing manual effort.

Primary Problem

High manual effort in claims review due to document complexity and repetitive data extraction.

Target Workflow

Document analysis and triage across PDFs, scanned forms, emails, and supporting evidence.

AI Capabilities

- **Classification:** Identify claim type, required workflow, and routing path.
- **Extraction:** Pull key data fields from unstructured documents with high accuracy.

Benefits

- Lower operational costs
- Faster claims processing
- Reduced backlog and rework
- Higher data accuracy and consistency
- Improved analyst productivity and morale

Required Guardrails

- Human-in-the-loop validation
- Confidence scoring and exception handling

- Audit trails for extracted data
- Bias and error monitoring
- Clear escalation paths for ambiguous cases

Expected Outcomes

- 30–70% reduction in manual review time
- Faster cycle times and improved SLAs
- More predictable operational costs
- Higher accuracy in claim classification and data capture

Workbook Exercise: Applying the Framework to Your Enterprise Workflow

Objective

Help managers translate the Mission → Problem → Workflow → AI Capability pattern into their own operational context.

Exercise 1 — Define the Mission

Write a one-sentence mission focused on operational efficiency or cost reduction.

Prompt: "What outcome reduces cost while improving service quality?"

Exercise 2 — Identify the Core Problem

Describe the operational barrier preventing the mission from being achieved.

Prompt: "Where is manual effort highest, and what makes it costly?"

Exercise 3 — Map the Workflow

Break the workflow into 5–7 steps. Identify where time, effort, or inconsistency is highest.

Prompt: "Which steps are repetitive, rules-driven, or document-heavy?"

Exercise 4 — Match AI Capabilities

Select the AI capability that best reduces the workflow friction.

Options include: classification, extraction, summarization, generation, prediction, recommendation, and anomaly detection.

Prompt: "Which capability automates the most expensive steps without replacing judgment?"

Exercise 5 — Validate Fit

Use these questions to confirm viability:

- Is the workflow high-volume or high-cost?
- Is the data available and accessible?
- Can humans review AI output?
- Are risks manageable with guardrails?

Exercise 6 — Draft Your Case Structure

Fill in your own version:

- **Mission:**
- **Problem:**
- **Workflow:**
- **AI Capability:**

4.13 Case Walkthrough: Enterprise Operations.

Case Study: Enterprise — Reducing Claims Review Costs with AI-Driven Document Intelligence

Mission: Reduce Operational Costs

Reducing operational costs is a central mission for enterprise organizations seeking to remain competitive while improving service quality. Cost reduction is not simply about cutting expenses—it is about eliminating inefficiencies, reducing rework, and reallocating human expertise to higher-value activities. When framed correctly, this mission encourages leaders to identify processes where manual effort is high, variability is costly, and automation can create a measurable financial impact. It also ensures that AI investments are evaluated not by novelty but by their ability to streamline operations, reduce cycle times, and improve the accuracy and consistency of core business processes.

Problem: High Manual Effort in Claims Review

Claims review is one of the most labor-intensive processes in many enterprise environments, particularly in insurance, financial services, and warranty operations. Analysts must sift through large volumes of documents—forms, receipts, reports, correspondence—to determine eligibility, completeness, and next steps. This manual effort increases operational costs, slows processing times, and introduces variability in decision quality. The problem is not only the volume of claims but the complexity and inconsistency of the documents themselves. By articulating the problem as excessive manual review effort, leaders can clearly see how document intelligence and automation can reduce cost drivers while improving throughput and accuracy.

Workflow: Document Analysis and Triage

Document analysis and triage is a structured yet highly repetitive workflow that begins when a claim is submitted and ends when it is routed to the appropriate team or decision path. Analysts must extract key data points, validate supporting evidence, classify claim types, and identify missing or inconsistent information. This workflow is ideal for AI augmentation because it involves predictable patterns, high document volume, and well-defined decision rules. Mapping the workflow reveals friction points, including manual data extraction, inconsistent classification, and delays due to incomplete submissions. These steps consume significant analyst time and create bottlenecks that directly increase operational costs.

AI Capability: Classification + Extraction

Classification and extraction capabilities allow AI systems to automatically categorize claims, identify claim types, and extract key data fields from unstructured documents. These models can process large volumes of PDFs, images, and forms, turning them into structured data that accelerates triage and reduces manual review. Classification models determine the claim category or required workflow, while extraction models pull out essential details such as dates, amounts, identifiers, and supporting evidence. When paired with human oversight, auditability, and domain-specific tuning, these capabilities significantly reduce manual effort, improve accuracy, and shorten processing times—directly supporting the mission of reducing operational costs.

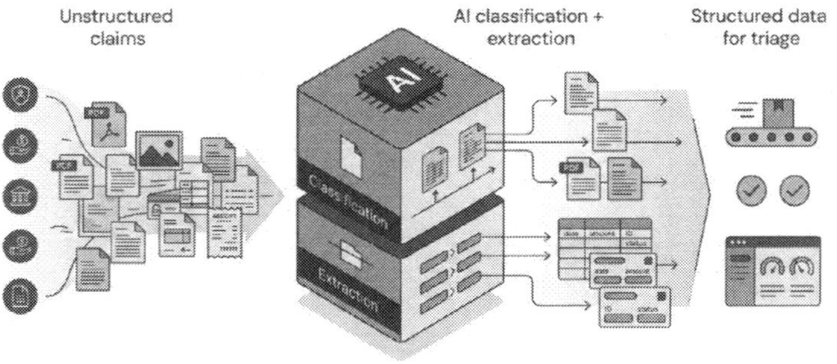

Full Case-Study Page

Enterprise Case Study: AI-Enabled Claims Review Automation

Context

Enterprises face rising operational costs driven by high document volume, complex claims, and manual review processes. Claims review is a prime candidate for AI augmentation because it is repetitive, rules-driven, and dependent on accurate document interpretation. This case study demonstrates how the Mission → Problem → Workflow → AI Capability framework guides leaders toward targeted, responsible AI adoption.

Mission

Reduce operational costs by streamlining claims review and reallocating analyst time to higher-value activities.

Problem

High manual effort in claims review due to document complexity, inconsistent formats, and repetitive data extraction.

Workflow

Document analysis and triage, including classification, data extraction, validation, and routing.

AI Capability

Classification + extraction to automatically categorize claims and extract key data fields from unstructured documents.

One-Page Handout: AI for Claims Review Automation

Overview

A concise reference for enterprise leaders evaluating AI opportunities in claims processing and document-heavy workflows.

Mission

Reduce operational costs by accelerating claims review and minimizing manual effort.

Primary Problem

High manual effort in claims review due to document complexity and repetitive data extraction.

Target Workflow

Document analysis and triage across PDFs, scanned forms, emails, and supporting evidence.

AI Capabilities

- **Classification:** Identify claim type, required workflow, and routing path.
- **Extraction:** Pull key data fields from unstructured documents with high accuracy.

Benefits

- Lower operational costs
- Faster claims processing
- Reduced backlog and rework
- Higher data accuracy and consistency
- Improved analyst productivity and morale

Required Guardrails

- Human-in-the-loop validation
- Confidence scoring and exception handling
- Audit trails for extracted data
- Bias and error monitoring
- Clear escalation paths for ambiguous cases

Expected Outcomes

- 30–70% reduction in manual review time
- Faster cycle times and improved SLAs
- More predictable operational costs
- Higher accuracy in claim classification and data capture

Workbook Exercise: Applying the Framework to Your Enterprise Workflow

Objective

Help managers translate the Mission → Problem → Workflow → AI Capability pattern into their own operational context.

Exercise 1 — Define the Mission

Write a one-sentence mission focused on operational efficiency or cost reduction.

Prompt: "What outcome reduces cost while improving service quality?"

Exercise 2 — Identify the Core Problem

Describe the operational barrier preventing the mission from being achieved.

Prompt: "Where is manual effort highest, and what makes it costly?"

Exercise 3 — Map the Workflow

Break the workflow into 5–7 steps. Identify where time, effort, or inconsistency is highest.

Prompt: "Which steps are repetitive, rules-driven, or document-heavy?"

Exercise 4 — Match AI Capabilities

Select the AI capability that best reduces the workflow friction.

Options include: classification, extraction, summarization, generation, prediction, recommendation, and anomaly detection.

Prompt: "Which capability automates the most expensive steps without replacing judgment?"

Exercise 5 — Validate Fit

Use these questions to confirm viability:

- Is the workflow high-volume or high-cost?
- Is the data available and accessible?
- Can humans review AI output?
- Are risks manageable with guardrails?

Exercise 6 — Draft Your Case Structure

Fill in your own version:

- **Mission:**
- **Problem:**
- **Workflow:**
- **AI Capability:**

4.14 Responsible AI Lens: Mission Protects People

Mission alignment is not just strategic — it is ethical.

It ensures:

- AI is used where it adds value
- AI does not replace essential human judgment
- AI does not harm vulnerable populations
- AI supports, rather than undermines, the mission
- AI is deployed with clarity and accountability

Mission is the first safeguard in responsible AI.

4.15 Summary — Key Takeaways

- AI strategy must begin with mission, not technology.
- The Mission → Problem → Workflow → AI chain is the foundation of responsible AI adoption.
- Mission alignment prevents wasted investment, fragmentation, and risk.
- AI is a workflow technology — if you cannot point to the workflow, you cannot apply AI effectively.
- Mission alignment is both a strategic and ethical requirement.

4.16 Workbook 4: Start With Mission, Not Models

This workbook section helps you apply the Mission → Problem → Workflow → AI framework to your organization.

Exercise 1 — Define Your Mission Outcomes

List the top 3–5 mission-critical outcomes for your organization.

Mission Outcome	Why It Matters	Priority (High/Med/Low)

Exercise 2 — Identify the Problems Blocking Each Mission Outcome

For each mission outcome, identify the problems preventing success.

Mission Outcome	Problem	Notes

Exercise 3 — Map Problems to Workflows

Identify where the problems actually occur.

Problem	Workflow	Who Performs It	Pain Points

Exercise 4 — Apply the Mission → Problem → Workflow → AI Chain

Choose one mission outcome and complete the chain.

Mission:

____

Problem:

____

Workflow:

AI Capability (if applicable):

Exercise 5 — Mission Alignment Scorecard

Rate each potential AI idea on mission alignment.

AI Idea	Mission Alignment (1–5)	Workflow Clarity (1–5)	Problem Clarity (1–5)	Notes

Reflection:

Ideas scoring below 3 on mission alignment should not move forward.

Diagram: Mission Alignment Scorecard

	A high-alignment opportunity is one in which the mission, the workflow, and the underlying problem all point in the same direction, giving teams the clarity and confidence to move quickly. When an initiative scores in this range, it signals strong strategic fit, well-understood processes, and a clearly defined need—conditions that dramatically increase the likelihood of meaningful impact.

	Medium-alignment opportunities often have a meaningful connection to the mission but lack the clarity needed to move forward with confidence. The intent is right, yet the workflow remains fuzzy—leaving teams unsure how the work will actually get done or how impact will be measured. This is the moment to pause, refine, and tighten the problem statement and process so the effort can either graduate into a high-alignment priority or be reshaped before resources are committed.
	Low-alignment opportunities signal that the work is too far removed from the mission and lacks the foundational clarity needed to succeed. When both the purpose and the workflow are unclear, teams risk investing time and resources into efforts that won't meaningfully advance strategic goals. Treat this score as a clear indicator to stop, reassess, and redirect energy toward initiatives with stronger alignment and a clearer path to impact.

5 Readiness: Data, People, Processes, Culture

Note that chapter 5 is a pivotal moment in the guide because it forces leaders to confront a simple truth: even the best AI strategy will fail if the organization isn't ready to support it. This chapter introduces the AI readiness assessment framework, providing managers with a structured way to evaluate the four dimensions that determine whether an initiative can take root: data, people, processes, and culture. It clarifies what "good enough" data maturity looks like (rather than perfection), how to gauge workforce readiness and change tolerance, and how to identify process complexity that either enables or blocks automation. By grounding readiness in observable conditions rather than optimism or assumptions, the chapter helps leaders avoid the costly mistake of pushing AI into environments that cannot sustain it.

Just as importantly, Chapter 5 shines a light on culture—the silent killer of AI initiatives. Even when data is available, teams are trained, and workflows are mapped, cultural resistance can quietly derail progress through fear, skepticism, or misaligned incentives. This chapter helps managers diagnose those cultural fault lines early and build strategies to address them before they become roadblocks. By the end, leaders gain a realistic picture of where their organization stands today, which gaps must be closed, and how to sequence AI efforts to match their true level of readiness.

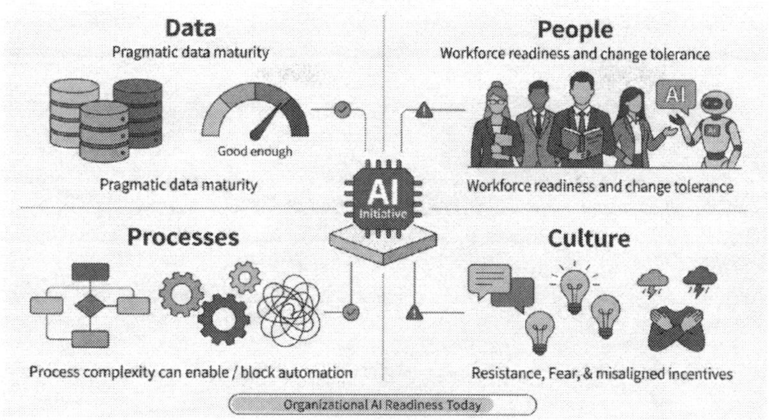

5.1 Opening Scenario: The Hidden Barrier

Your leadership team is aligned. Your mission priorities are clear. Your AI value opportunities look promising.

Then you try to implement your first pilot — and everything slows down.

- The data isn't clean enough
- The workflow isn't standardized
- The team is overwhelmed
- Compliance raises concerns
- IT says the infrastructure isn't ready
- Staff are anxious about job impact

This is the moment when many AI initiatives stall. Not because the strategy was wrong — but because the organization wasn't ready.

AI readiness is not a technical assessment. It is an organizational one.

This chapter gives you a practical, manager-friendly framework to assess readiness across four dimensions:

1. **Data**
2. **People**
3. **Processes**
4. **Culture**

These dimensions determine whether your AI strategy can move from idea to execution.

5.2 Why Readiness Matters More Than Technology

The model rarely limits AI success.

It is limited by the environment in which the model must operate.

Organizations that skip readiness assessment experience:

- Delays
- Rework
- Resistance
- Compliance issues
- Poor adoption
- Failed pilots

Organizations that assess readiness early experience:

- Faster implementation
- Higher adoption
- Lower risk
- Clearer expectations
- Better outcomes

Readiness is not a gate. It is a roadmap.

5.3 The Four Dimensions of AI Readiness

Here is the framework you will use throughout the rest of the book.

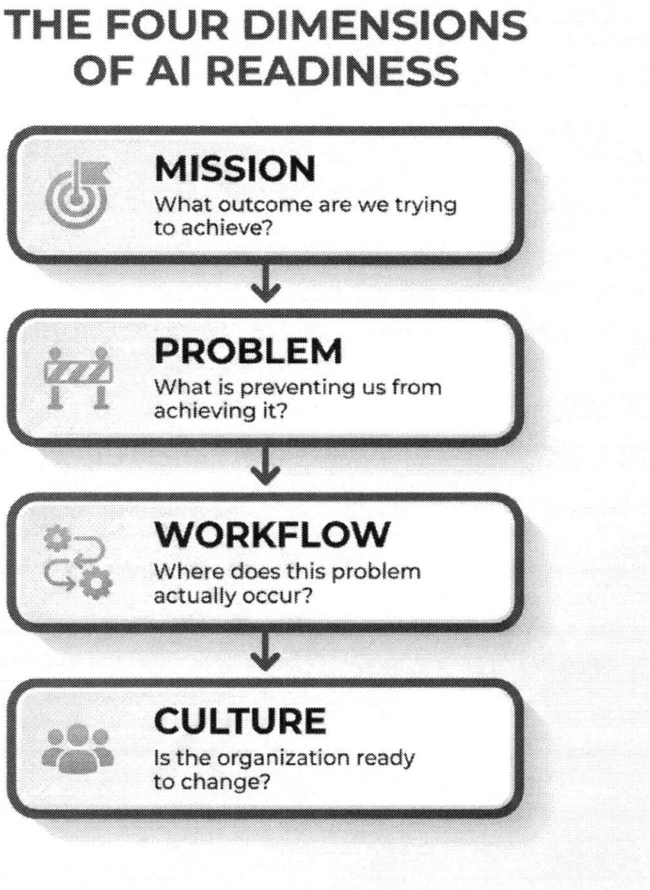

Each dimension must be assessed honestly and pragmatically.

1. Data Readiness: Is the Data Usable?

AI does not require perfect data. But it does require **usable** data.

Managers should assess data readiness using four questions:

1. Do we know where the data lives?

If data is scattered across systems, AI will struggle.

2. Is the data accessible?

If access requires months of approvals, pilots will stall.

3. Is the data structured enough?

AI can work with unstructured data — but only if it's consistent.

4. Is the data accurate enough?

AI amplifies whatever it is given.

Manager's rule: You don't need perfect data. You need *fit-for-purpose* data.

2. People Readiness: Are Teams Prepared to Adopt AI?

AI changes how people work. People's readiness is often the biggest barrier.

Assess people's readiness by asking:

1. Do teams understand the purpose of AI?

If not, they will resist.

2. Do they trust technology?

Trust must be earned through transparency.

3. Do they have the skills to use AI tools?

Training is not optional.

4. Do they have the capacity to participate in pilots?

AI requires time — for testing, feedback, and iteration.

Manager's rule: AI adoption is a human transformation, not a technical one.

3. Process Readiness: Are Workflows Stable Enough for AI?

AI cannot fix broken workflows. It can only accelerate them.

Assess process readiness by asking:

1. Is the workflow documented?

If not, AI cannot be inserted reliably.

2. Is the workflow consistent across teams?

Variation kills automation.

3. Are the inputs and outputs predictable?

AI thrives on consistency.

4. Are there clear decision points?

AI needs defined boundaries.

Manager's rule: If the workflow is chaotic, fix the workflow before adding AI.

4. Culture Readiness: Is the Organization Ready for Change?

Culture determines whether AI is embraced or rejected.

Assess culture readiness by asking:

1. Is leadership aligned and supportive?

Mixed messages create confusion.

2. Is there psychological safety?

People must feel safe experimenting.

3. Is there openness to new ways of working?

Rigid cultures resist AI.

4. Is there a history of successful change?

Past behavior predicts future adoption.

Manager's rule: Culture is the multiplier — it amplifies success or failure.

Diagram: The Four Dimensions of AI Readiness

How to Conduct a Readiness Assessment

Use a simple three-step process:

Step 1: Score Each Dimension (1–5)

1 = very low readiness5 = very high readiness

Step 2: Identify Gaps

Ask:

- What is slowing us down?
- What is creating risk?
- What must be fixed before pilots begin?

Step 3: Build a Readiness Action Plan

For each gap, define:

- What needs to change
- Who owns the change
- How long will it take
- What support is required

This becomes part of your 90-day roadmap in Chapter 8.

Case Example: Public Sector Readiness

- **Data:** Scattered across legacy systems → Score: 2
- **People:** High motivation, low training → Score: 3
- **Processes:** Highly variable → Score: 2
- **Culture:** Strong mission alignment → Score: 4

Conclusion: Fix processes and data access before launching pilots.

Case Example: Healthcare Readiness

- **Data:** High volume, inconsistent quality → Score: 3
- **People:** Burnout risk, limited capacity → Score: 2
- **Processes:** Documented but inefficient → Score: 3
- **Culture:** Open to innovation → Score: 4

Conclusion: Start with low-burden pilots that reduce workload.

5.4 Responsible AI Lens: Readiness Protects People

Readiness assessment is a safety mechanism.

It ensures:

- AI is deployed where it can succeed
- People are not harmed by rushed implementation
- Workflows are stable enough to support oversight
- Data is accurate enough to avoid bias
- Culture is strong enough to support responsible use

Responsible AI begins with readiness.

5.5 Summary — Key Takeaways

- AI readiness determines whether your strategy can be executed.
- Readiness has four dimensions: data, people, processes, and culture.
- You don't need perfect data — you need usable data.
- People's readiness is often the biggest barrier.
- AI cannot fix broken workflows.
- Culture is the multiplier of success or failure.
- Readiness assessment is essential for responsible AI adoption.

5.6 Companion Workbook 5: Assessing Readiness

This workbook section helps you evaluate your organization's readiness across the four dimensions.

Exercise 1 — Readiness Scoring

Score each dimension from 1–5.

Table 9 Readiness Assessment

Dimension	Score (1-5)	Notes
Data	☐	
People	☐	
Processes	☐	
Culture	☐	

Reflection:

Which dimension is your biggest barrier?

Exercise 2 — Data Readiness Checklist

Check all that apply:

- ☐ We know where the data lives
- ☐ We can access the data
- ☐ The data is structured enough
- ☐ The data is accurate enough
- ☐ The data is consistent across teams
- ☐ We have metadata or documentation

If fewer than 3 boxes are checked: Data readiness is a risk.

Exercise 3 — People Readiness Assessment

Rate agreement (1 = strongly disagree, 5 = strongly agree):

Table 10 Team Evaluation Table

Statement	Score
Teams understand why we are adopting AI.	
Teams Trust the Technology	

Teams have the capacity to participate.	

Exercise 4 — Process Stability Scan

Identify workflows that are unstable.

Table 11 Stability Assessment

Workflow	Stability (1-5)	Notes

Table 12 Workflow Risk Evaluation Table

Workflow	Volume (1-5)	Variability (1-5)	Risk (1-5)	Notes

Exercise 5 — Culture Readiness Reflection

Answer the prompts:

- How does leadership talk about AI?
- How does staff feel about AI?
- What past change efforts succeeded or failed?
- What cultural barriers exist?

____

____

____

Exercise 6 — Readiness Action Plan

For each low-scoring dimension, define actions.

Dimension	Gap	Action Needed	Owner	Timeline

6 Identifying and Prioritizing Use Cases

Note that chapter 6 is where AI strategy becomes operationally real. Up to this point, you've built a foundation of understanding, alignment, and readiness—but Chapter 6 is where you turn that groundwork into a concrete portfolio of AI opportunities. The five-step use case discovery workshop gives leaders a structured, repeatable way to surface real problems worth solving, grounded in mission and workflow rather than abstract brainstorming. By using the Use Case Scorecard—evaluating impact, feasibility, risk, and time to value—you gain a disciplined method for comparing ideas that may look equally promising on the surface but differ dramatically in their likelihood of success. This chapter ensures that prioritization is not driven by enthusiasm, politics, or vendor pressure, but by clear, transparent criteria that everyone can understand.

Just as importantly, Chapter 6 teaches you how to run a cross-functional prioritization session that brings together operations, IT, data, compliance, and frontline teams. This is where alignment is forged, and blind spots are surfaced early. The chapter also shows how to build a balanced AI portfolio that mixes quick wins—projects that build momentum and confidence—with strategic bets that advance long-term transformation. Without this balance, organizations either move too slowly or overreach and stall. By the end of Chapter 6, you have a clear, defensible, and mission-aligned set of AI initiatives that your organization can rally around and execute with confidence.

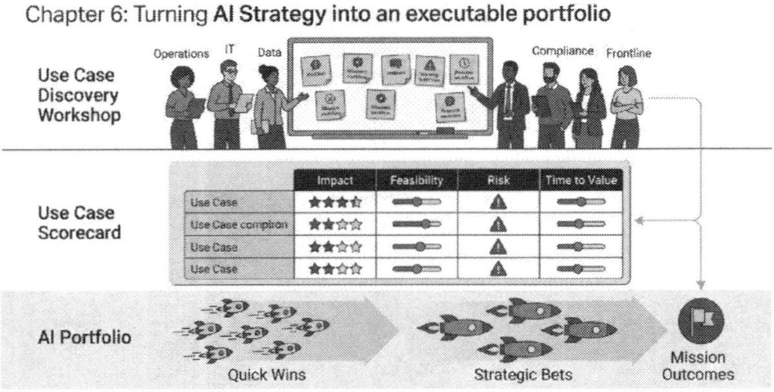

From ideas to a balanced, mission-aligned AI portfolio

6.1 Scenario: Too Many Ideas, Not Enough Clarity

Your inbox fills with AI ideas:

- "Can we automate this?"
- "Can AI summarize that?"
- "Can we use AI to predict this?"
- "Can we build a chatbot for that?"

Everyone has ideas. Most of them are interesting. Some of them are valuable. Very few of them are strategic.

This is the moment where AI strategies succeed or fail.

The problem isn't a lack of ideas — it's a lack of prioritization.

This chapter gives you a structured, repeatable process to identify, evaluate, and prioritize AI use cases that are:

- Mission-aligned
- High-value
- Feasible

- Responsible
- Scalable

This is where your AI strategy comes to life.

6.2 Why Use Case Prioritization Matters

Organizations that skip prioritization end up with:

- Random pilots
- Low-impact wins
- Fragmented solutions
- Wasted investment
- Shadow AI
- Loss of credibility

Organizations that prioritize effectively achieve:

- Clear focus
- High-value outcomes
- Faster wins
- Better adoption
- Lower risk
- Stronger executive support

Prioritization is not about choosing what's exciting. It's about choosing what matters.

6.3 The Five-Step Use Case Discovery Workshop

This is the workshop you will run with your teams. It is simple, structured, and highly effective.

1. Mission Alignment

2. Workflow Mapping

3. Pain Point Identification

4. AI Opportunity Mapping

5. Prioritization

Let's break it down.

Step 1 — Mission Alignment

Start with the mission outcomes from Chapter 4.

Ask:

- What outcomes matter most this year?
- What outcomes are underperforming?
- What outcomes have the highest strategic importance?

Use cases that do not support mission outcomes should not move forward.

Step 2 — Workflow Mapping

AI is a workflow technology. You must identify the workflow before identifying the use case.

Ask:

- Where does the work actually happen?
- Who performs it?
- What steps are involved?
- What data is used?
- Where are the bottlenecks?

Workflows reveal opportunities.

Step 3 — Pain Point Identification

Pain points are where AI creates value.

Common pain points include:

- High manual effort
- Long cycle times
- High error rates
- High variability
- High backlog
- High cognitive load
- High administrative burden

Pain points are the raw material for use cases.

Step 4 — AI Opportunity Mapping

Now — and only now — do you ask:

"Can AI help?"

Use the four AI capability categories from Chapter 2:

- Classification
- Prediction
- Generation
- Reasoning

Match capabilities to pain points.

Step 5 — Prioritization

Use the **Use Case Scorecard** to evaluate each idea.

The scorecard evaluates:

- **Value**
- **Feasibility**
- **Risk**
- **Impact**
- **Readiness**

This creates a clear, defensible prioritization.

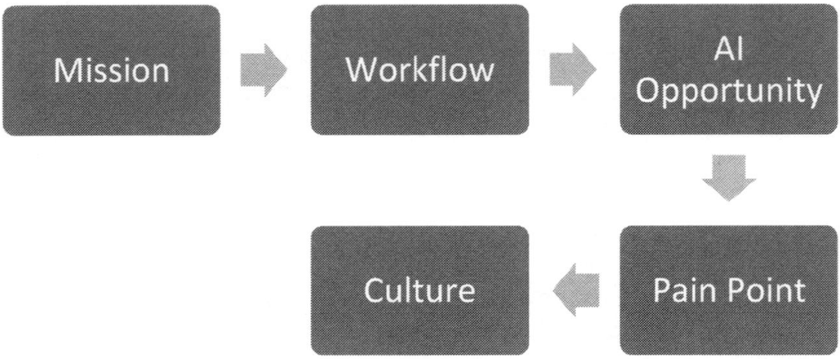

Figure 4 Use Case Discovery Process

6.4 The Use Case Scorecard

This is the core tool for prioritization.

Each use case is scored 1–5 on five dimensions:

1. Value

- Efficiency
- Effectiveness
- Insight

2. Feasibility

- Data availability
- Workflow consistency
- Technical complexity

3. Risk

- Impact of errors
- Regulatory constraints

• Safety considerations

4. Impact

• Number of people affected
• Frequency of workflow
• Mission importance

5. Readiness

• Data readiness
• People readiness
• Process readiness
• Culture readiness

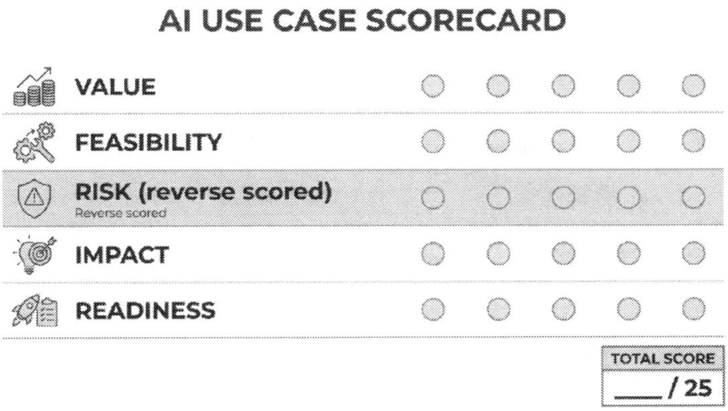

How to Prioritize Use Cases

Use this simple rule:

High Priority

• High value
• High impact
• Medium–high feasibility

- Low–medium risk
- Medium–high readiness

Medium Priority

- High value but low readiness
- High impact but high risk
- Medium value but high feasibility

Low Priority

- Low value
- Low impact
- High risk
- Low readiness

Do Not Pursue

- No mission alignment
- No clear workflow
- No data
- High risk + low value

Case Example: Healthcare

Use Case: Clinical note summarization

- **Value**: High
- **Feasibility**: High
- **Risk**: Medium
- **Impact**: High
- **Readiness**: Medium
- **Priority**: High

Case Example: Public Sector

Use Case: Citizen inquiry routing

- **Value**: High
- **Feasibility**: High
- **Risk**: Low
- **Impact**: High
- **Readiness**: High
- **Priority**: Very High

Case Example: Enterprise

- **Use Case:** Predictive maintenance
- **Value:** High
- **Feasibility:** Medium
- **Risk:** Medium
- **Impact:** High
- **Readiness:** Low
- **Priority:** Medium (requires readiness work)

The Responsible AI Lens: Prioritization Protects People

Prioritization ensures:

- AI is used where it is safe
- High-risk workflows receive oversight
- Vulnerable populations are protected
- AI is deployed responsibly
- Resources are used wisely

Responsible AI is not a separate step — it is built into the prioritization process.

6.5 Summary — Key Takeaways

- Use case discovery must be structured, not ad hoc.
- Mission alignment is the first filter.
- Workflows reveal where AI can help.
- Pain points are the raw material for use cases.

- The Use Case Scorecard creates clarity and defensibility.
- Prioritization ensures responsible, high-value AI adoption.

Workbook 6 — Identifying & Prioritizing Use Cases

This workbook helps you run the Use Case Discovery Workshop.

Exercise 1 — Mission-Aligned Use Case Brainstorm

List mission outcomes and brainstorm use cases.

Mission Outcome	Potential Use Cases	Notes

Exercise 2 — Workflow Mapping

Choose one workflow and map it.

Workflow Name:

____

Steps:

1. ____
2. ____
3. ____
4. ____

Pain Points:

- ____
- ____

•

Exercise 3 — AI Opportunity Mapping

Match pain points to AI capabilities.

Pain Point	AI Capability	Notes
	Classification	
	Prediction	
	Generation	
	Reasoning	

Exercise 4 — Use Case Scorecard

Score each use case 1–5.

Use Case	Value	Feasibility	Risk*	Impact	Readiness	Total

*Risk is reverse-scored.

Exercise 5 — Prioritization Matrix

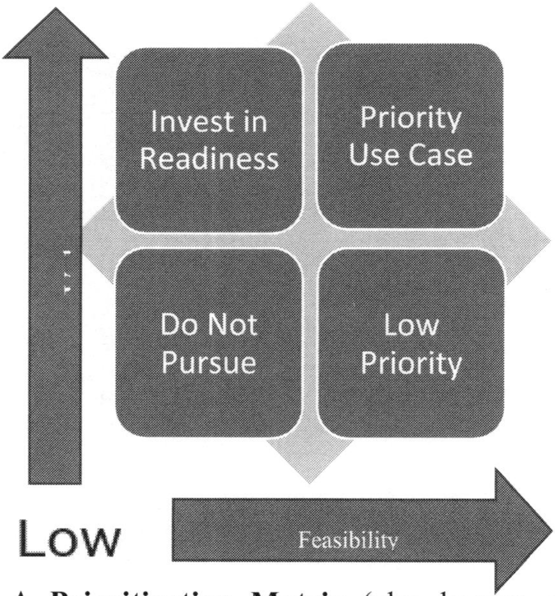

A **Prioritization Matrix** (also known as an Impact vs. Feasibility Matrix) is a strategic tool used to categorize tasks or AI use cases by balancing their potential value against the technical or operational ease of implementing them.

The Four Strategic Quadrants

- **Priority Use Cases (High Value | High Feasibility):** These are your "Quick Wins." They offer significant benefits and are easy to execute, making them the top candidates for immediate action.
- **Invest in Readiness (High Value | Low Feasibility):** These represent "Strategic Bets." While the value is high, current resources or technology may not be ready. You should allocate resources to bridge these gaps for future implementation.
- **Low Priority (Low Value | High Feasibility):** Often called "Fill-ins" or "Busy Work," these tasks are easy to do but provide little ROI. They should only be addressed if time and capacity allow.
- **Do Not Pursue (Low Value | Low Feasibility):** These are "Resource Drains" or "Money Pits." They offer minimal impact and require high effort; they should be removed from your backlog entirely.

THE USE CASE DISCOVERY BLUEPRINT

1. Mission Alignment

- What outcome are we trying to achieve?

2. Workflow Mapping

- Where does the work actually happen?

3. Pain Point Identification

- What slows us down?

4. AI Opportunity Mapping

- Which AI capability fits the pain point?

5. Prioritization

- Use the Scorecard to evaluate:
- Value
- Feasibility
- Risk
- Impact
- Readiness

Output

A ranked list of high-value, mission-aligned, feasible AI use cases.

7 Designing the AI Operating Model

Note that chapter 7 is essential because it defines the organizational backbone that determines whether AI efforts scale smoothly or collapse under their own weight. This chapter helps leaders understand the tradeoffs between centralized, federated, and hybrid AI team structures—and why the hybrid model is emerging as the most resilient for complex organizations. It clarifies the distinct roles of product managers, data scientists, engineers, and domain experts, showing how each contributes to building AI systems that are not only technically sound but operationally relevant. By making these roles explicit, Chapter 7 helps leaders avoid the common pitfall of assuming "the data team will figure it out," and instead builds a shared understanding of who owns what across the AI lifecycle.

Just as importantly, Chapter 7 introduces governance structures that prevent chaos without strangling innovation. As AI initiatives multiply, organizations often struggle with duplicated work, inconsistent standards, unclear decision rights, and unmanaged risk. This chapter provides a blueprint for responsible scaling—how to set guardrails, define approval pathways, and maintain quality while still empowering teams to move quickly. It shows leaders how to build an operating model that can absorb growth, adapt to new technologies, and maintain alignment across business units. By the end of Chapter 7, managers have a clear picture of how to structure their teams, processes, and governance to enable AI to scale responsibly, sustainably, and with real impact.

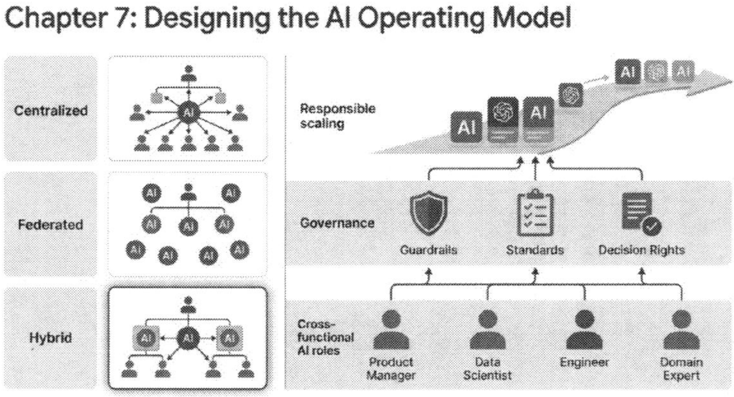

7.1 Opening Scenario: The Scaling Problem

Your first AI pilot succeeds. Leaders are excited. Teams want more. Demand explodes.

Then reality hits:

- Who owns AI across the organization?
- Who approves of new use cases?
- Who maintains models?
- Who ensures responsible AI compliance?
- Who trains staff?
- Who manages vendors?
- Who decides what gets built next?

Without a clear operating model, AI becomes chaotic. Teams duplicate work. Risk increases. Pilots stall. Momentum dies.

This chapter gives you the structure to prevent that chaos — and to scale AI responsibly and sustainably.

7.2 What Is an AI Operating Model?

An AI operating model defines **how your organization will run AI:**

- Who makes decisions
- Who owns what
- How workflows execute
- How risk is managed
- How teams collaborate
- How AI is monitored
- How AI is scaled

It is the **organizational blueprint** for AI adoption.

Without it, AI becomes a collection of disconnected experiments.

With it, AI becomes a strategic capability.

7.3 The Need for AI Oper. Model Before Scaling

Organizations that skip the experience of operating modeling:

- Confusion about roles
- Slow decision-making
- Shadow AI
- Inconsistent standards
- Compliance issues
- Low adoption
- High risk

Organizations that define the operating model early experience:

- Faster scaling
- Clear accountability
- Better governance
- Higher adoption
- Lower risk
- More predictable outcomes

The operating model is the bridge between strategy and execution.

7.4 The Three AI Operating Model Archetypes

There are three common ways organizations structure AI teams:

- **Centralized**
- **Federated**
- **Hybrid**

Each has strengths and weaknesses.

1. Centralized Operating Model

All AI capabilities reside in a single central team.

Strengths

- Strong governance
- Consistent standards
- Efficient resource use
- Clear accountability

Weaknesses

- Can become a bottleneck

- Limited domain expertise
- Slower to respond to local needs

Best for

- Early-stage AI adoption
- High-risk environments
- Organizations with limited AI talent

2. Federated Operating Model

AI capabilities are distributed across business units.

Strengths

- Deep domain expertise
- Faster innovation
- High ownership

Weaknesses

- Inconsistent standards
- Higher risk
- Duplication of effort
- Harder to govern

Best for

- Mature AI organizations
- Large enterprises
- Highly diverse business units

3. Hybrid Operating Model (Recommended)

A central team sets standards for governance and shared tools. Business units build and deploy AI with support.

Strengths

- Balance of speed and control
- Shared infrastructure
- Consistent governance
- Domain-specific innovation

Weaknesses

- Requires strong coordination
- Needs clear roles and responsibilities

Best for

- Most organizations
- Especially the public sector, healthcare, and enterprise

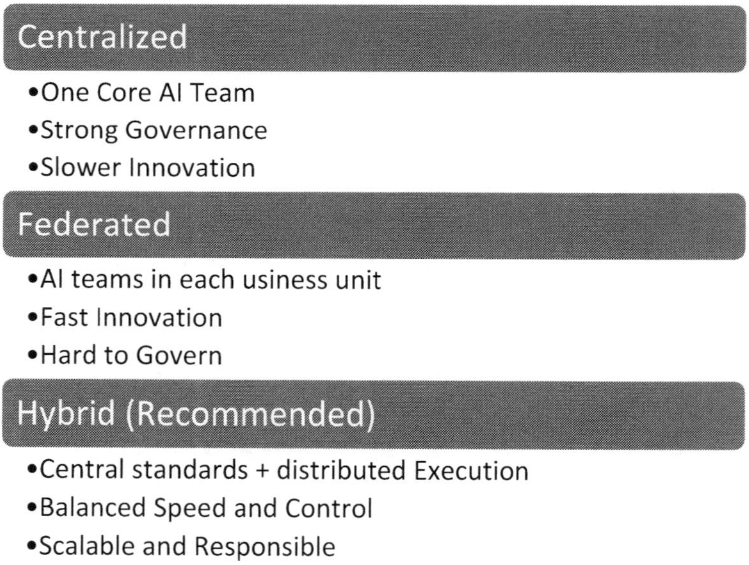

Diagram: The Three Operating Model Archetypes

7.5 The Roles Needed in Your AI Operating Model

Regardless of structure, every AI operating model needs these roles:

1. AI Strategy Lead

Owns the AI roadmap and aligns AI to mission outcomes.

2. Product Owner / Workflow Owner

Owns the workflow AI is improving. Ensures adoption and success.

3. Data Owner

Ensures data access, quality, and governance.

4. Responsible AI Lead

Ensures compliance, fairness, transparency, and safety.

5. Technical Lead

Oversees model selection, integration, and performance.

6. Change Management Lead

Ensures training, communication, and adoption.

7. Executive Sponsor

Provides authority, funding, and strategic alignment.

7.6 Governance: The Backbone of the Oper. Model

Governance ensures AI is:

- Safe
- Responsible
- Compliant
- Transparent
- Monitored
- Auditable

Governance should include:

- Use case intake and prioritization
- Risk assessment
- Human-in-the-loop design
- Model evaluation
- Monitoring and drift detection
- Incident response
- Documentation standards

Governance is not bureaucracy. It is protection — for people, for the mission, and for the organization.

7.7 How to Choose the Right Operating Model

Use these three questions:

1. How mature is your AI capability?

- Low maturity → centralized
- High maturity → federated
- Growing maturity → hybrid

2. How high is the risk environment?

High risk → centralized or hybrid

Low risk → federated

3. How diverse are your workflows?

Highly diverse → hybrid

Homogeneous → centralized

Case Example: Public Sector (Hybrid Model)

- Central team sets standards
- Agencies build their own solutions

- Shared governance ensures safety
- Shared infrastructure reduces cost

Case Example: Healthcare (Centralized → Hybrid)

- Start centralized for safety
- Move to hybrid as maturity grows
- Clinical workflows require strong oversight

Case Example: Enterprise (Federated → Hybrid)

- Business units innovate quickly
- Central team adds governance and shared tools
- Hybrid becomes the long-term model

The Responsible AI Lens: Operating Models Protect People

A strong operating model ensures:

- AI is used responsibly
- Risks are identified early
- Oversight is clear
- Accountability is defined
- Vulnerable populations are protected
- Decisions are transparent

Responsible AI is not a separate process.
It is built into the operating model.

7.8 Summary — Key Takeaways

- The AI operating model defines how AI is run across the organization.
- Without it, AI becomes chaotic and risky.

- There are three archetypes: centralized, federated, and hybrid.
- Hybrid is the most scalable and is responsible for most organizations.
- Clear roles are essential for success.
- Governance is the backbone of responsible AI.
- The operating model must evolve as maturity grows.

7.9 Workbook 7: Designing the AI Oper. Model

This workbook helps you design the operating model that fits your organization.

Exercise 1 — Identify Your Current Operating Model

Check the model that best describes your current state:

- ☐ Centralized
- ☐ Federated
- ☐ Hybrid
- ☐ Undefined/unclear

Notes:

____

Exercise 2 — Operating Model Fit Assessment

Rate each model on fit (1–5):

Model	Fit Score (1–5)	Notes
Centralized		
Federated		
Hybrid		

Exercise 3 — Define Your AI Roles

List who currently fills each role (or who should).

Role	Current Owner	Gaps / Needs
AI Strategy Lead		
Product Owner		
Data Owner		
Responsible AI Lead		
Technical Lead		
Change Mgmt. Lead		
Executive Sponsor		

Exercise 4 — Governance Checklist

Check all that exists today:

- ☐ Use case intake process
- ☐ Prioritization criteria
- ☐ Responsible AI review
- ☐ Human-in-the-loop design
- ☐ Model evaluation standards
- ☐ Monitoring and drift detection
- ☐ Incident response process
- ☐ Documentation standards

If fewer than 4 are checked: Governance is a risk.

Exercise 5 — Operating Model Action Plan

Define the steps needed to move to your target model.

Area	Current State	Target State	Actions Needed	Owner
Roles				
Governance				
Infrastructure				
Processes				
Culture				

Part III – Roadmapping for Execution

Part III, *Roadmapping for Execution*, turns strategy into movement. This section focuses on how to convert your mission-aligned AI priorities into a clear, sequenced, and realistic plan that teams can actually deliver. It walks you through defining 30--, 60--, and 90-day milestones; identifying dependencies; mapping workflows; and determining what needs to be built, bought, or integrated. Instead of vague aspirations or disconnected pilots, Part III helps you create a roadmap that shows exactly what will happen, when, and why—anchored in operational reality rather than wishful thinking. This is where AI stops being a concept and becomes a coordinated effort across people, processes, data, and technology.

Just as importantly, this section teaches leaders how to manage risk, pace, and expectations. AI initiatives often fail not because the strategy was wrong, but because execution lacked structure: unclear ownership, unrealistic timelines, missing prerequisites, or poor sequencing. Part III provides the tools to avoid those pitfalls by helping you define roles, establish governance, set measurable checkpoints, and build feedback loops that keep the roadmap adaptive. By the end, you have a plan that is not only actionable but resilient—capable of surviving shifting priorities, technical surprises, and organizational constraints while still driving meaningful progress.

8 Building the 90-Day AI Roadmap

Note that chapter 8 is where strategy becomes momentum. After defining your mission, identifying use cases, and assessing readiness, this chapter shows why a **90-day window** is the ideal planning horizon for AI execution. Ninety days is long enough to make meaningful progress—establishing foundations, launching pilots, and preparing for scale—yet short enough to maintain urgency, visibility, and cross-functional alignment. This chapter breaks the work into three clear horizons—**foundation, pilots, and scale**—so leaders can see how each phase builds on the last. Instead of vague timelines or open-ended initiatives, you gain a structured, time-boxed approach that keeps teams focused and accountable.

Just as importantly, Chapter 8 teaches you how to **sequence work to avoid bottlenecks**, one of the most common reasons AI projects stall. By mapping dependencies across data, people, processes, and technology, leaders learn how to prevent classic pitfalls such as waiting on data access, unclear ownership, misaligned expectations, or pilots that launch before prerequisites are ready. The chapter also provides 90-day roadmap templates you can adapt to your organization, giving you a practical starting point rather than a blank page. By the end, you have a realistic roadmap — mission-aligned and execution-ready—one that builds confidence, accelerates learning, and sets the stage for sustainable scaling.

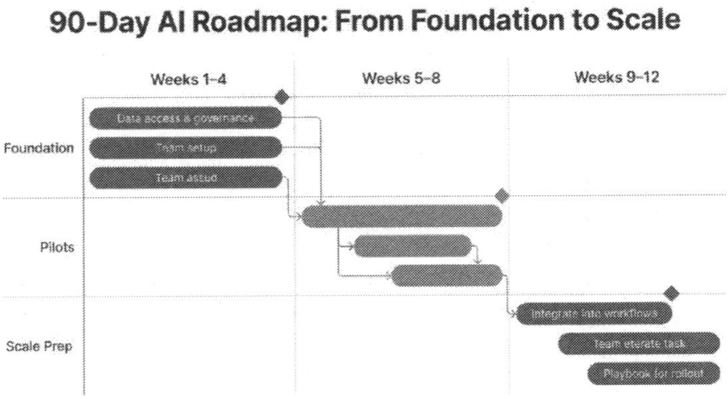

8.1 Opening Scenario: The Momentum Problem

You've aligned on the mission. You've identified high-value use cases. You've assessed readiness. You've designed your operating model.

Now leadership asks the question that determines whether your AI strategy becomes real:

"What are we doing in the next 90 days?"

This is where many organizations freeze. Not because they lack ideas — but because they lack structure.

A 90-day roadmap solves this problem. It creates:

- Focus
- Momentum
- Accountability
- Measurable progress
- Executive confidence
- Organizational clarity

This chapter gives you the framework to build a 90-day roadmap that is realistic, mission-aligned, and execution-ready.

8.2 Why 90 Days Is the Sweet Spot

90 days is long enough to make meaningful progress, but short enough to:

- Maintain urgency
- Avoid overplanning
- Reduce risk
- Keep teams aligned
- Adapt to new information
- Deliver visible wins

Annual plans are too slow. Weekly plans are too tactical. 90 days is the strategic middle ground.

8.3 The Three Horizons of the 90-Day Roadmap

Every successful AI roadmap includes three horizons:

Horizon 1 → Foundation

Horizon 2 → Pilots

Horizon 3 → Scale Preparation

Let's break them down.

Horizon 1 — Foundation (Weeks 1–4)

This is where you set the conditions for success.

Key Activities

- Finalize use case selection
- Define success metrics

- Confirm data access
- Establish governance checkpoints
- Assign roles and responsibilities
- Conduct risk assessment
- Begin workflow documentation
- Prepare training and change management plans

Output

- Use case charter
- Data access plan
- Responsible AI checklist
- Workflow map
- Success metrics

Manager's rule: If you skip the foundation, the pilot will fail.

Horizon 2 — Pilots (Weeks 5–10)

This is where you test, learn, and refine.

Key Activities

- Build or configure the AI solution
- Integrate into the workflow
- Conduct user testing
- Collect feedback
- Measure performance
- Adjust prompts, models, or workflows
- Document risks and mitigations

Output

- Working pilot
- User feedback summary

- Performance dashboard
- Updated workflow
- Risk mitigation plan

Manager's rule: Pilots are experiments — not commitments.

Horizon 3 — Scale Preparation (Weeks 11–13)

This is where you decide whether to scale, refine, or stop.

Key Activities

- Evaluate pilot results
- Assess adoption readiness
- Identify scaling requirements
- Estimate cost and resource needs
- Document lessons learned
- Prepare executive recommendation

Outputs

- Scale/no-scale decision
- Scaling plan
- Updated roadmap
- Executive briefing

Manager's rule: Scaling is a business decision, not a technical one.

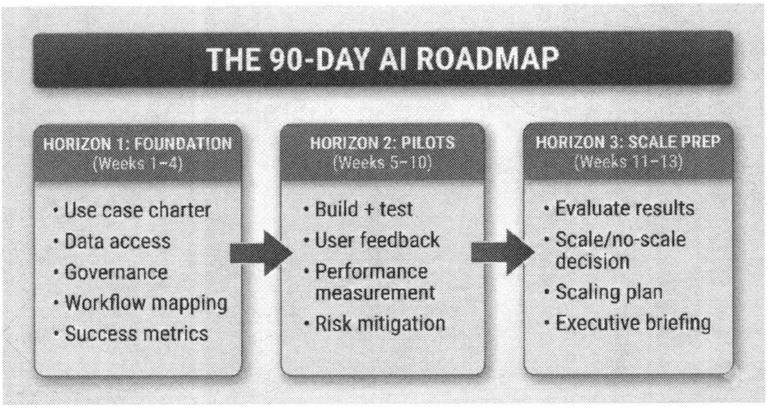

8.4 The Use Case Charter (Your Roadmap Anchor)

Every roadmap begins with a **Use Case Charter**.

It includes:

- Mission alignment
- Workflow description
- Success metrics
- Data requirements
- Responsible AI considerations
- Roles and responsibilities
- Timeline
- Risks and mitigations

This becomes the single source of truth for the next 90 days.

8.5 Success Metrics: What Good Looks Like

Success metrics must be:

- Clear
- Measurable
- Mission-aligned
- Time-bound
- Realistic

Examples:

- Reduce documentation time by 30%
- Improve triage accuracy by 15%
- Reduce backlog by 20%
- Improve citizen response time by 40%
- Reduce manual review by 50%

Metrics create accountability and clarity.

8.6 The Pilot-to-Production Pathway

Pilots are not the end. They are the beginning.

The pathway looks like this:

Scaling requires:

- Stable workflows
- Clear governance
- Strong adoption
- Reliable performance
- Documented risks
- Executive support

Scaling is not automatic — it is earned.

Case Example: Healthcare

- **Use Case:** Clinical note summarization
- **Goal:** Reduce documentation time by 25%
- **Pilot Result:** 32% reduction, high clinician satisfaction
- **Decision:** Scale to additional departments
- **Case Example: Public Sector**
- **Use Case:** Citizen inquiry routing
- **Goal:** Reduce response time by 40%
- **Pilot Result:** 55% reduction, low risk
- **Decision:** Scale immediately

Case Example: Enterprise

- **Use Case:** Predictive maintenance
- **Goal:** Reduce downtime by 20%
- **Pilot Result:** Inconclusive due to data issues
- **Decision:** Extend pilot after data cleanup

The Responsible AI Lens: Roadmaps Reduce Risk

A 90-day roadmap ensures:

- AI is deployed responsibly
- Risks are identified early
- Oversight is built in
- Pilots are safe and controlled
- Scaling decisions are evidence-based

Responsible AI is not a separate process — it is embedded in the roadmap.

Summary — Key Takeaways

- A 90-day roadmap creates momentum and clarity.
- The roadmap has three horizons: foundation, pilots, and scale prep.

- The Use Case Charter anchors the roadmap.
- Success metrics must be clear and measurable.
- Pilots are experiments — not commitments.
- Scaling requires evidence, readiness, and governance.
- Roadmaps are essential for responsible AI adoption.

8.7 Workbook 8: Building the 90-Day AI Roadmap

This workbook helps you build your first 90-day roadmap.

Exercise 1 — Define Your Use Case Charter

Use Case Name:

Mission Alignment:

Workflow:

Success Metrics:

- ______________________________
- ______________________________
- ______________________________

Data Requirements:

- ______________________________
- ______________________________

Risks & Mitigations:

- ______________________________
- ______________________________

Exercise 2 — Horizon 1 Planning (Weeks 1–4)

List your foundation tasks.

Task	Owner	Due Date

Exercise 3 — Horizon 2 Planning (Weeks 5–10)

List your pilot tasks.

Task	Owner	Due Date

Exercise 4 — Horizon 3 Planning (Weeks 11–13)

List your scale-prep tasks.

Task	Owner	Due Date

Exercise 5 — Pilot Evaluation Checklist

Check all that apply:

- ☐ Success metrics met
- ☐ Users satisfied
- ☐ Risks mitigated
- ☐ Workflow stable
- ☐ Data reliable
- ☐ Governance followed
- ☐ Adoption readiness high

Scale Decision:

- ☐ Scale

- ☐ Refine
- ☐ Stop

9 Running AI Pilots That Actually Work

Note that chapter 9 is critical because it teaches managers how to run AI pilots that generate real evidence—not just activity. Many organizations launch pilots that look promising on paper but fail to produce measurable outcomes, clear learnings, or a path to scale. This chapter breaks down the anatomy of a successful pilot, showing how to define the problem precisely, set the right scope, secure the necessary data and workflows, and establish success metrics that matter. It helps leaders avoid the common trap of treating pilots as experiments with no accountability, instead positioning them as structured tests designed to validate assumptions, reduce uncertainty, and build organizational confidence.

Equally important, Chapter 9 addresses the phenomenon of "pilot purgatory," in which projects linger indefinitely without being shut down or scaled back. This chapter gives managers a disciplined framework for deciding when to kill a pilot, pivot, and invest in scaling. It emphasizes that ending a pilot is not failure—it's responsible leadership when the evidence doesn't support further investment. By the end of Chapter 9, leaders know how to design pilots that produce actionable insights, make data-driven decisions about next steps, and ensure that only the strongest, most mission-aligned initiatives move forward into full deployment.

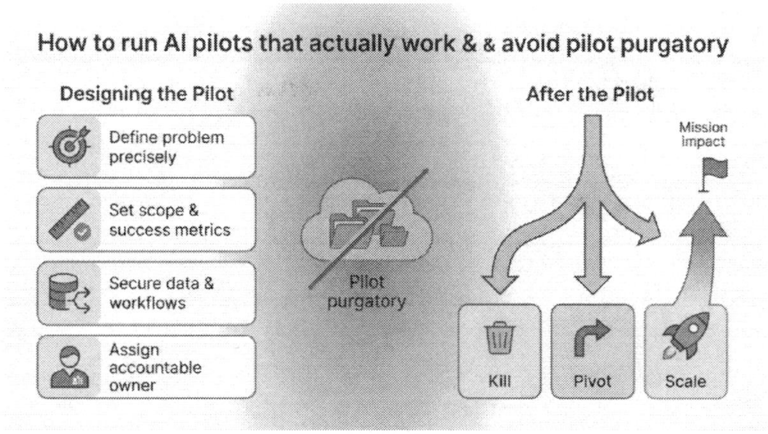

9.1 Opening Scenario: The Pilot Trap

Your team launches an AI pilot.

- The demo looks great.
- Leadership is excited.
- The vendor is confident.
- The slide deck is polished.

Then the real-world hits:

- Users don't adopt it
- The workflow doesn't fit
- The data isn't consistent
- The model behaves unpredictably
- Compliance raises concerns
- IT says it can't scale
- The pilot never leaves the sandbox

This is the pilot trap — where most AI initiatives go to die.

This chapter shows you how to run pilots that:

- Deliver real value
- Build trust
- Reduce risk
- Generate evidence
- Prepare for scale
- Avoid "pilot purgatory."

This is where your AI strategy becomes operational reality.

9.2 Difference of AI Pilot vs. Traditional Pilot

AI pilots are not like software pilots. They require a different mindset.

AI pilots are:

- Iterative
- Data-dependent
- Behavior-driven
- Workflow-sensitive
- Risk-aware
- Human-in-the-loop
- Evidence-based

Traditional pilots are:

- Feature-driven
- Static
- Pass/fail
- Technology-centric

AI pilots are experiments — not deployments.

9.3 The Anatomy of a Successful AI Pilot

A successful AI pilot has **six essential components**:

1. Clear problem definition

2. Defined success metrics

3. Stable workflow

4. Human-in-the-loop design

5. Responsible AI safeguards

6. Continuous measurement and iteration

Let's break these down.

1. Clear Problem Definition

The pilot must solve a real, documented problem.

Ask:

- What pain point are we addressing?
- Who experiences it?
- How often does it occur?
- What is the current cost or burden?
- If the problem is unclear, the pilot will fail.

2. Defined Success Metrics

Success must be measurable.

Examples:

- Reduce documentation time by 30%
- Improve routing accuracy by 20%
- Reduce backlog by 40%
- Improve customer satisfaction by 15%
- Reduce manual review by 50%

Metrics create accountability and clarity.

3. Stable Workflow

AI cannot fix broken workflows. It can only accelerate them.

Before piloting, ensure:

- The workflow is documented
- The workflow is consistent
- Inputs and outputs are predictable
- Decision points are clear
- If the workflow is unstable, fix it first.

4. Human-in-the-Loop Design

AI does NOT replace Humans — it augments them.

Human-in-the-loop design ensures:

- Safety
- Accuracy
- Trust
- Accountability

Examples:

- Humans approve AI-generated summaries
- Humans validate AI-classified cases
- Humans override AI recommendations

This is essential in healthcare, the public sector, and enterprise environments.

5. Responsible AI Safeguards

Every pilot must include:

- Bias assessment
- Privacy review
- Security controls
- Transparency requirements
- Error handling
- Auditability

Responsible AI is not optional — it is foundational.

6. Continuous Measurement and Iteration

AI pilots are not “set it and forget it.”

You must:

- Monitor performance
- Collect user feedback
- Adjust prompts or models
- Refine workflows
- Document issues
- Re-evaluate risks

Iteration is the engine of improvement.

Anatomy of a Successful Project

CLEAR PROBLEM –
A specific, well-defined operational problem with measurable stakes.

SUCCESS METRICS –
Mission-aligned metrics established before the pilot begins.

STABLE WORKFLOW –
A consistent, repeatable workflow so the pilot evaluates AI—not variability.

HUMAN-IN-THE-LOOP –
Defined oversight points to ensure safety, quality, and accountability.

RESPONSIBLE AI SAFEGUARDS –
Guardrails for privacy, bias, compliance, and security.

9.4 The Pilot Lifecycle

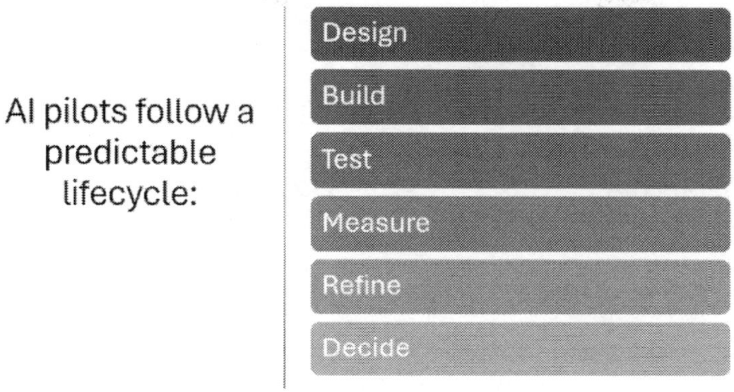

Let's walk through each stage.

Stage 1 — Design

Define:

- Problem
- Workflow
- Success metrics
- Data requirements
- Risks
- Human oversight
- Governance checkpoints

This is the most important stage.

Stage 2 — Build

Activities include:

- Model selection
- Prompt engineering
- Workflow integration
- UI/UX design
- Data access configuration

Build only what is needed for the pilot — not the full solution.

Stage 3 — Test

Test with real users in real workflows.

Collect:

- Accuracy data
- Error patterns
- User feedback
- Workflow friction points

Testing reveals reality.

Stage 4 — Measure

Measure against your success metrics.

Ask:

- Did we reduce time?
- Did we improve accuracy?
- Did we reduce burden?
- Did users trust the system?
- Did risks emerge?

Measurement determines truth.

Stage 5 — Refine

Refine:

- Prompts
- Models
- Workflows
- UI
- Safeguards

Iteration is where value emerges.

Stage 6 — Decide

Make an evidence-based decision:

- Scale
- Refine and extend
- Stop

Scaling is not automatic — it is earned.

9.5 How to Avoid Pilot Purgatory

Pilot purgatory happens when:

- No success metrics
- No owner
- No timeline
- No governance
- No decision criteria
- No executive visibility

To avoid it:

- Set a 90-day limit
- Define success metrics
- Assign a pilot owner
- Establish governance checkpoints
- Require an executive decision at the end

Pilots must end — Not Drift.

Case Example: Healthcare

Use Case: Clinical note summarization

Pilot Result:

- 32% reduction in documentation time
- High clinician satisfaction
- Low risk

Decision: Scale

Case Example: Public Sector

Use Case: Citizen inquiry routing

Pilot Result:

- 55% reduction in response time
- High accuracy
- Strong adoption

Decision: Scale immediately

Case Example: Enterprise

Use Case: Predictive maintenance

Pilot Result:

- Inconsistent data
- Low accuracy
- High variability

Decision: Extend pilot after data cleanup

The Responsible AI Lens: Pilots Are Safety Mechanisms

Pilots ensure:

- AI is safe
- AI is fair
- AI is transparent
- AI is accurate
- AI is trustworthy
- AI is aligned with the mission

Pilots protect people — and the organization.

9.6 Summary — Key Takeaways

- AI pilots are experiments, not deployments.
- Successful pilots require clear problem definition, metrics, workflows, oversight, safeguards, and iteration.
- The pilot lifecycle is: design → build → test → measure → refine → decide.
- Human-in-the-loop design is essential.
- Responsible AI must be embedded from the start.
- Pilots must end with a scale/no-scale decision.

9.7 Workbook 9: Running AI Pilots That Work

This workbook helps you design and run a successful AI pilot.

Exercise 1 — Pilot Definition

Use Case:

____

Problem:

____

Workflow:

____

Exercise 2 — Success Metrics

List 3–5 measurable success metrics.

1. ____

2. ____

3. ____

4.____

Exercise 3 — Human-in-the-Loop Design

Define human oversight.

AI Does:

- ____
- ____

Humans Do:

- ____
- ____

Exercise 4 — Responsible AI Safeguards

Check all that apply:

- ☐ Bias assessment
- ☐ Privacy review
- ☐ Security controls
- ☐ Error handling
- ☐ Transparency requirements
- ☐ Auditability
- ☐ Human override

Exercise 5 — Pilot Lifecycle Plan

Stage	Tasks	Owner	Due Date
Design			
Build			

Test			
Measure			
Refine			
Decide			

Exercise 6 — Pilot Evaluation Checklist

Check all that apply:

- ☐ Success metrics met
- ☐ Users satisfied
- ☐ Risks mitigated
- ☐ Workflow stable
- ☐ Data reliable
- ☐ Governance followed
- ☐ Adoption readiness high

Decision:

- ☐ Scale
- ☐ Refine
- ☐ Stop

10 Scaling AI Across the Enterprise

Note that chapter 10 matters because it marks the transition from *successful pilots* to *enterprise-level transformation*. At this point, AI stops being a series of experiments and becomes a durable capability embedded across the organization. This chapter introduces the **AI flywheel**, a model that explains how early wins generate data, build trust, and create reusable components that accelerate every subsequent deployment. It also breaks down the three dominant scaling patterns—**horizontal**, **vertical**, and **workflow-based**—so leaders can choose the right expansion path rather than defaulting to ad hoc growth. By making scaling intentional rather than opportunistic, Chapter 10 helps organizations avoid the common trap of scattered, one-off solutions that never add up to meaningful impact.

Just as importantly, Chapter 10 shows how to build the **reusable components**—data pipelines, model templates, governance patterns, integration layers—that turn AI from a craft project into an operational engine. It also addresses the rising risks that come with scale: model drift, compounding errors, privacy exposure, and inconsistent oversight. Leaders learn how to manage these risks without slowing innovation, using governance structures that grow with the organization rather than constrain it. By the end of the chapter, readers will understand how to scale AI responsibly, sustainably, and strategically, creating an enterprise where each new deployment is faster, safer, and more valuable than the last.

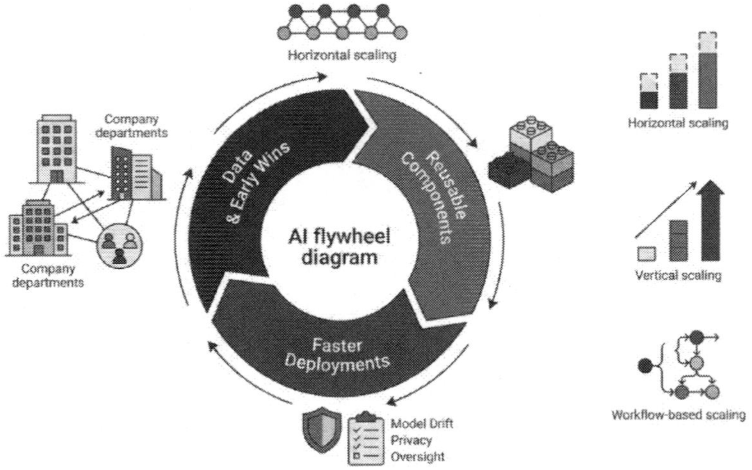

10.1 Opening Scenario: The Scaling Challenge

Your first pilots succeed.

- Leaders are excited.
- Teams want more.
- Demand skyrockets.

But scaling AI is not the same as launching a pilot.

Suddenly, you face new questions:

- How do we scale without breaking workflows?
- How do we maintain quality and safety?
- How do we avoid duplicating work across teams?
- How do we ensure consistent governance?
- How do we manage risk at scale?
- How do we build reusable components?
- How do we keep momentum without losing control?

Scaling AI is not about doing more pilots. It is about building a **repeatable system** that turns successful pilots into enterprise-wide capabilities.

This chapter gives you the blueprint.

10.2 Why Scaling Is Harder Than Piloting

Pilots are controlled environments. Scaling is the real world.

Scaling introduces:

- More users
- More workflows
- More variability
- More data
- More risk
- More oversight
- More integration
- More complexity

Most organizations underestimate this shift — and their AI programs stall.

Scaling requires **structure**, **governance**, **infrastructure**, and **change management**.

10.3 The AI Flywheel: How Scaling Really Works

Scaling AI is not linear — it is a **flywheel**.

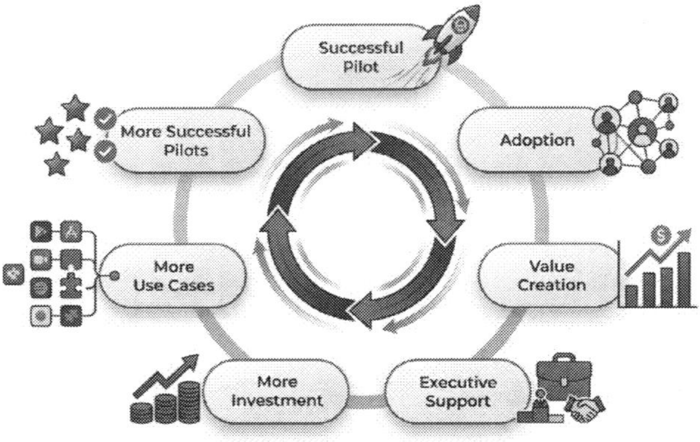

The flywheel accelerates when:

- Pilots deliver measurable value
- Users adopt the solution
- Leaders see results
- Investment increases
- More use cases enter the pipeline

Your job is to keep the flywheel spinning.

10.4 The Three Scaling Patterns

Organizations scale AI in three primary ways:

- **Horizontal Scaling**
- **Vertical Scaling**
- **Workflow Scaling**

Let's break them down.

---

1. Horizontal Scaling (Across Teams)

You deploy the same AI capability across multiple teams or departments.

Examples:

- Document summarization across all business units
- Inquiry routing across all service centers
- Knowledge search across all departments

Strengths

- Fast
- High ROI
- Reusable components

Risks

- Inconsistent adoption
- Workflow variation

---

2. Vertical Scaling (Deeper in One Domain)

You expand AI within a single domain or function.

Examples:

- Scaling AI in clinical documentation across specialties
- Scaling AI in fraud detection across programs
- Scaling AI in supply chain across product lines

Strengths

- Deep expertise
- Strong adoption
- High quality

Risks

- Siloed innovation
- Limited cross-team reuse

---

3. Workflow Scaling (End-to-End)

You expand AI across an entire workflow, not just a single task.

Examples:

- Intake → triage → routing → summarization
- Detection → investigation → adjudication
- Draft → review → approval → communication

Strengths

- Transformational impact
- Major efficiency gains

Risks

- High complexity
- Requires strong governance

Diagram: The Three Scaling Patterns

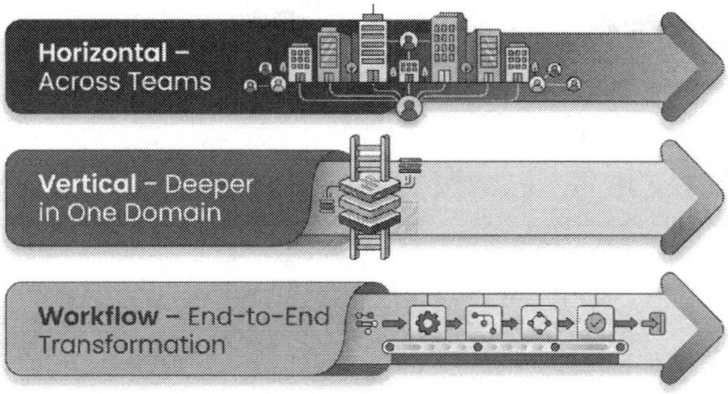

Building Reusable AI Components

Scaling requires **reuse**, not reinvention.

Reusable components include:

- Prompt libraries
- Model templates
- Workflow patterns
- Data pipelines
- Evaluation frameworks
- Governance checklists
- Monitoring dashboards

Reuse accelerates scaling and reduces risk.

Governance at Scale

Scaling increases risk — and governance must evolve.

Enterprise-scale governance includes:

- Standardized intake

- Centralized risk assessment
- Model registries
- Monitoring and drift detection
- Audit logs
- Human-in-the-loop requirements
- Incident response plans

Governance is not a blocker — it is an enabler.

Change Management at Scale

Scaling AI requires:

- Training
- Communication
- Role clarity
- Incentives
- Support channels
- Champions networks

People don't resist AI — they resist unclear expectations.

The Scale Readiness Checklist

Before scaling, ensure:

- ☐ Success metrics met
- ☐ Workflow stable
- ☐ Users satisfied
- ☐ Risks mitigated
- ☐ Governance approved
- ☐ Data reliable

- ☐ Adoption readiness high
- ☐ Infrastructure ready

If any box is unchecked, refine before scaling.

Case Example: Healthcare

- **Pilot:** Clinical note summarization
- **Scaling Pattern:** Horizontal
- **Result:** Expanded to all departments
- **Impact:** 30% reduction in documentation time

Case Example: Public Sector

- **Pilot:** Inquiry routing
- **Scaling Pattern:** Workflow
- **Result:** End-to-end transformation of citizen services
- **Impact:** 55% reduction in response time

Case Example: Enterprise

- **Pilot:** Predictive maintenance
- **Scaling Pattern:** Vertical
- **Result:** Expanded across product lines
- **Impact:** 20% reduction in downtime

The Responsible AI Lens: Scaling Increases Risk

Scaling amplifies:

- Bias
- Errors
- Privacy concerns
- Security risks
- Operational impact

Responsible AI must scale with the solution.

10.5 Summary — Key Takeaways

- Scaling AI is harder than piloting — and far more valuable.
- The AI Flywheel accelerates when pilots deliver value.
- There are three scaling patterns: horizontal, vertical, and workflow.
- Reusable components accelerate scaling.
- Governance must evolve as scale increases.
- Change management is essential for adoption.
- Scaling requires readiness, evidence, and oversight.

10.6 Workbook 10: Scaling AI Across the Enterprise

This workbook helps you design your scaling strategy.

Exercise 1 — Identify Your Scaling Pattern

Check the pattern that fits your use case:

- ☐ Horizontal
- ☐ Vertical
- ☐ Workflow

Ask why this pattern fits:

____

Exercise 2 — Scale Readiness Assessment

Check all that apply:

- ☐ Success metrics met
- ☐ Workflow stable
- ☐ Users satisfied
- ☐ Risks mitigated
- ☐ Governance approved
- ☐ Data reliable
- ☐ Adoption readiness high
- ☐ Infrastructure ready

If fewer than 6 boxes are checked: Scaling is not yet recommended.

Exercise 3 — Reusable Components Inventory

List reusable components you can leverage.

Component	Exists? (Y/N)	Notes
Prompt library		
Model template		
Workflow pattern		

Component	Exists? (Y/N)	Notes
Data pipeline		
Evaluation framework		
Governance checklist		
Monitoring dashboard		

Exercise 4 — Scaling Plan

Area	Actions Needed	Owner	Timeline
Governance			
Infrastructure			
Training			
Communication			
Workflow updates			
Risk mitigation			

Exercise 5 — Executive Scaling Brief

Draft your scaling recommendation.

Summary:

____

Evidence:

____

Risks:

____

Recommendation:

- ☐ Scale
- ☐ Refine
- ☐ Stop

10.7 One-Page Handout: The AI Scaling Blueprint

Perfect for workshops, leadership briefings, or onboarding.

THE AI SCALING BLUEPRINT

1. Confirm Readiness

- Success metrics met
- Workflow stable
- Risks mitigated
- Governance approved

2. Choose Scaling Pattern

- Horizontal
- Vertical
- Workflow

3. Build Reusable Components

- Prompts
- Models
- Pipelines
- Governance
- Monitoring

4. Strengthen Governance

- Risk assessment
- Model registry
- Drift detection
- Audit logs

5. Drive Adoption

- Training
- Communication
- Champions
- Support

6. Scale Responsibly

- Monitor
- Evaluation
- Adjust
- Improve

Part IV Responsible, Safe, and Sustainable AI

Part IV brings the book's most important shift: moving from *building AI* to *building AI that is responsible, safe, and sustainable at scale*. This section explains why responsible AI is not just an ethical obligation but a competitive advantage—strengthening trust, reducing operational risk, and enabling faster adoption across the enterprise. It introduces the six pillars of responsible AI. It shows how to embed them directly into workflows, so safety becomes part of everyday operations rather than an after-the-fact review. Managers get practical tools, checklists, and governance patterns that help them anticipate risks, document decisions, and create reusable safeguards that grow with the organization. Together, these chapters equip leaders to scale AI confidently, ensuring every deployment is fair, secure, transparent, and aligned with organizational values.

11 Responsible AI by Design

Note that chapter 11 is essential because it reframes responsible AI not as a compliance burden but as a **strategic advantage**. As organizations scale AI across workflows, the stakes rise: decisions become faster, automation touches more people, and risks—bias, privacy breaches, safety failures—can compound quickly. This chapter introduces the **six pillars of responsible AI**, providing leaders with a practical structure to ensure that every system they deploy is safe, fair, transparent, and aligned with organizational values. Rather than treating responsibility as an after-the-fact audit, Chapter 11 positions it as a design principle that strengthens trust, reduces operational risk, and accelerates adoption.

The chapter also shows managers how to **embed safety directly into workflows**, making responsible AI a natural part of day-to-day operations rather than a specialized function. It provides practical tools—checklists, decision frameworks, governance patterns—that help teams evaluate risks early, document decisions, and create consistent guardrails across the enterprise. By the end of Chapter 11, leaders understand how responsible AI enables scale, protects the organization, and builds durable confidence among employees, customers, and regulators.

11.1 Opening Scenario: The Responsibility Gap

Your AI pilot is ready. The workflow is mapped. The model performs well. The team is excited.

Then someone asks:

- "How do we know the model is fair?"
- "What happens if it makes a mistake?"
- "How do we explain its decisions?"
- "What data was it trained on?"
- "What if it drifts over time?"
- "Who is accountable if something goes wrong?"

Suddenly, the room gets quiet.

This is the responsibility gap — the space between what AI *can* do and what your organization is *prepared* to manage.

Responsible AI is not a compliance exercise.

It is a strategic advantage — and a requirement for safe, sustainable AI adoption.

This chapter provides the framework to embed responsible AI at every stage of your AI strategy.

11.2 Why Responsible AI Matters

AI introduces new risks:

- Bias
- Hallucinations
- Privacy violations
- Security vulnerabilities
- Misuse
- Over-reliance
- Lack of transparency
- Model drift

These risks are not theoretical — they are operational.

Responsible AI ensures:

- Safety
- Fairness
- Transparency
- Accountability
- Trust
- Compliance
- Mission alignment

Responsible AI is not a barrier to innovation.
It is the foundation that makes innovation possible.

11.3 The Six Pillars of Responsible AI

Every organization needs a clear, actionable framework. Here are the six pillars you will use throughout the rest of the book.

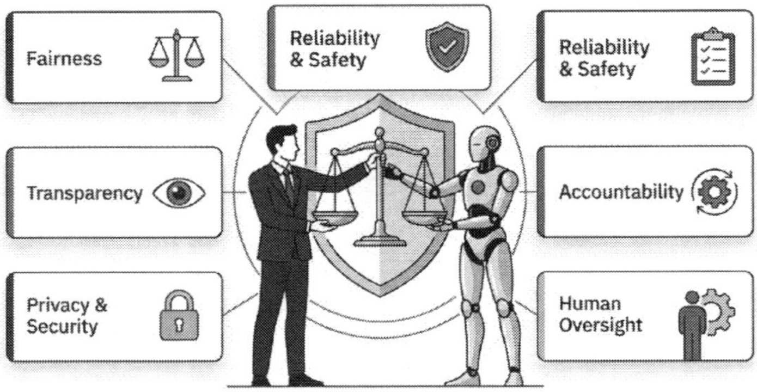

Let's break them down.

---

1. Fairness

AI must not create or amplify bias.

Ask:

- Who could be harmed?
- Are outcomes equitable across groups?
- Is the data representative?
- Are there historical biases in the workflow?

Fairness is not about perfection — it is about vigilance.

2. Reliability & Safety

AI must perform consistently and safely.

Ask:

- Does the model behave predictably?
- Does it hallucinate?
- How does it perform on edge cases?
- How will we monitor drift?

Safety is a continuous process.

---

3. Privacy & Security

AI must protect sensitive data.

Ask:

- What data is used?
- How is it stored?
- Who has access?
- How is it protected?
- What happens if there is a breach?

Privacy is non-negotiable.

---

4. Transparency

Users must understand how AI works — at the right level.

Ask:

- Can we explain how the model makes decisions?
- Can users understand its limitations?
- Can we document its behavior?

Transparency builds trust.

---

5. Accountability

Humans — not AI — are responsible for outcomes.

Ask:

- Who owns the workflow?
- Who approves AI decisions?
- Who monitors performance?
- Who responds to incidents?

Accountability must be explicit.

---

6. Human Oversight

AI augments humans — it does not replace them.

Ask:

- Where must humans stay in the loop?
- When can humans override AI?
- How do we ensure humans remain engaged?

Oversight is the ultimate safeguard.

11.4 Responsible AI by Design, Practical Framework

Responsible AI must be embedded into every stage of the AI lifecycle.

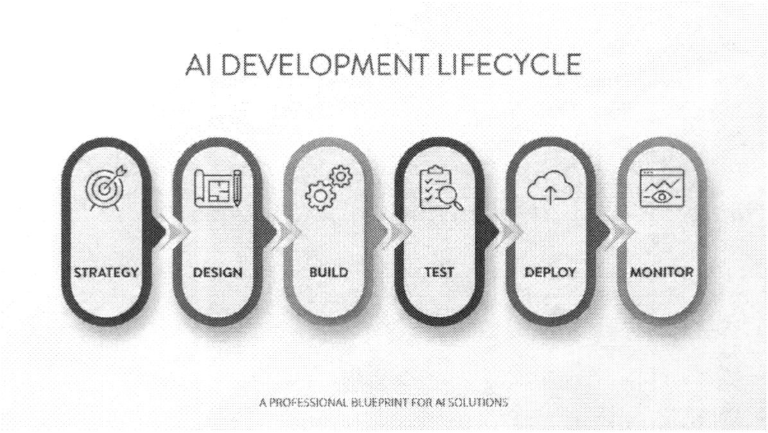

Let's walk through each stage.

---

1. Strategy

Define:

- Mission alignment
- Risks
- Impacted populations
- Ethical considerations

Responsible AI begins with intent.

---

2. Design

Design for:

- Human oversight
- Transparency
- Safety
- Privacy

Design choices determine risk.

---

3. Build

Implement:

- Guardrails
- Error handling
- Logging
- Access controls

Build with responsibility in mind.

---

4. Test

Test for:

- Bias
- Accuracy
- Safety
- Edge cases
- User experience

Testing reveals reality.

---

5. Deploy

Deploy with:

- Governance
- Documentation
- Training
- Monitoring plans

Deployment is not the finish line — it is the starting line.

---

6. Monitor

Monitor:

- Drift
- Errors
- User feedback
- Performance
- Compliance

Monitoring protects People and the Mission.

11.5 The Responsible AI Checklist

Use this checklist for every use case.

Fairness

- ☐ Data reviewed for bias
- ☐ Outcomes tested across groups

Safety

- ☐ Edge cases tested
- ☐ Hallucination risk assessed

Privacy

- ☐ Sensitive data protected
- ☐ Access controls in place

Transparency

- ☐ Model behavior documented
- ☐ User guidance provided

Accountability

- ☐ Workflow owner identified
- ☐ Incident response defined

Human Oversight

- ☐ Human-in-the-loop defined
- ☐ Override mechanisms in place

Case Example: Healthcare

Use Case: Clinical note summarization
Responsible AI Focus:

- Human oversight required
- Privacy controls are essential
- Transparency for clinicians

Case Example: Public Sector

Use Case: Eligibility screening

Responsible AI Focus:

- High fairness requirements
- High accountability
- Clear override mechanisms

Case Example: Enterprise

Use Case: Customer service automation

Responsible AI Focus:

- Transparency for customers
- Safety for edge cases
- Monitoring for drift

The Responsible AI Lens: Responsibility Is Leadership

Responsible AI is not a technical requirement — it is a leadership requirement.

It ensures:

- Trust
- Safety
- Mission alignment
- Ethical integrity
- Sustainable adoption

Responsible AI is how you lead in the age of AI.

Summary — Key Takeaways

- Responsible AI is a strategic advantage.
- The six pillars are fairness, safety, privacy, transparency, accountability, and oversight.
- Responsible AI must be embedded into every stage of the lifecycle.
- Checklists and governance reduce risk.
- Responsible AI protects people, the mission, and the organization.

11.6 Workbook 11 — Responsible AI by Design

This workbook helps you apply responsible AI principles to your use cases.

Exercise 1 — Responsible AI Risk Scan

Identify risks for your use case.

Risk Category	Risks Identified	Notes
Fairness		
Safety		
Privacy		
Transparency		
Accountability		
Oversight		

Exercise 2 — Human-in-the-Loop Design

Define oversight.

AI Does:

- ______________________________
- ______________________________

Humans Do:

- ____________________________________
- ____________________________________

Exercise 3 — Responsible AI Checklist

Check all that apply:

- ☐ Bias assessment completed
- ☐ Privacy review completed
- ☐ Safety testing completed
- ☐ Transparency documentation created
- ☐ Workflow owner assigned
- ☐ Human oversight defined
- ☐ Monitoring plan created

Exercise 4 — Responsible AI Action Plan

Area	Action Needed	Owner	Timeline
Fairness			
Safety			
Privacy			
Transparency			
Accountability			

Area	Action Needed	Owner	Timeline
Oversight			

One-Page Handout: The Responsible AI Blueprint

Perfect for workshops, leadership briefings, or onboarding.

THE RESPONSIBLE AI BLUEPRINT

The Six Pillars

- Fairness
- Reliability & Safety
- Privacy & Security
- Transparency
- Accountability
- Human Oversight

Lifecycle

Checklist

- Bias assessed
- Privacy protected
- Safety tested
- Transparency documented
- Accountability assigned
- Oversight defined

- Monitoring active

Principle

Responsible AI is not a constraint — it is a competitive advantage.

12 Managing Risk, Compliance, and Security

Note that chapter 12 is a turning point because it confronts the reality that AI at scale introduces new categories of organizational risk—operational, ethical, regulatory, and security-related—that leaders must manage deliberately rather than reactively. As AI systems become more deeply embedded in workflows, the consequences of failure grow: biased outputs can affect real people, model drift can compromise decisions, and opaque systems can expose organizations to compliance risks. This chapter breaks down the major **AI risk categories** and provides clear **mitigation strategies**, giving leaders a structured way to anticipate issues before they materialize. It also clarifies the evolving **regulatory landscape** across both the public sector and enterprise environments, helping organizations understand what is required, what is emerging, and how to embed compliance into their AI programs from day one.

The chapter also addresses the often-overlooked challenge of **vendor risk management**, recognizing that most organizations rely on a complex ecosystem of models, APIs, cloud services, and third-party tools. It explains how to evaluate vendors, assess shared responsibility models, and ensure contractual and technical safeguards are in place. Finally, it explores the **security implications of AI systems**, including data leakage, prompt injection, model exploitation, and the new attack surfaces created by autonomous agents and integrated workflows. By the end of Chapter 12, leaders understand how to build AI systems that are not only powerful but also secure, compliant, and resilient, ensuring that innovation does not outpace the organization's ability to govern it responsibly.

12.1 Opening Scenario: The Risk Reality Check

Your AI pilot is ready to scale. The workflow is stable. The model performs well. Users are excited.

Then the risk, compliance, and security teams join the meeting.

They ask:

- "What data does the model use?"
- "How do we prevent unauthorized access?"
- "What happens if the model makes a harmful mistake?"
- "How do we ensure compliance with regulations?"
- "How do we detect drift or misuse?"
- "What is our incident response plan?"

Suddenly, your timeline feels fragile.

This is not resistance — it is responsibility.

AI introduces new categories of risk that traditional IT systems never faced. Managing these risks is not optional. It is essential for safe, sustainable, mission-aligned AI adoption.

This chapter gives you the framework to do exactly that.

12.2 Why AI Risk Is Different

AI risk is fundamentally different from traditional software risk.

Traditional software risk is deterministic.

If the code is correct, the output is predictable.

AI risk is probabilistic.

Even a well-performing model can produce unexpected outputs.

AI introduces risks related to:

- Bias
- Hallucinations
- Privacy
- Security
- Misuse
- Drift
- Over-reliance
- Lack of transparency
- Regulatory exposure

These risks must be managed proactively — not reactively.

12.3 The Four Categories of AI Risk

Every AI risk falls into one of four categories:

1. Model Risk

2. Data Risk

3. Operational Risk

4. Compliance & Regulatory Risk

Let's break them down.

1. Model Risk

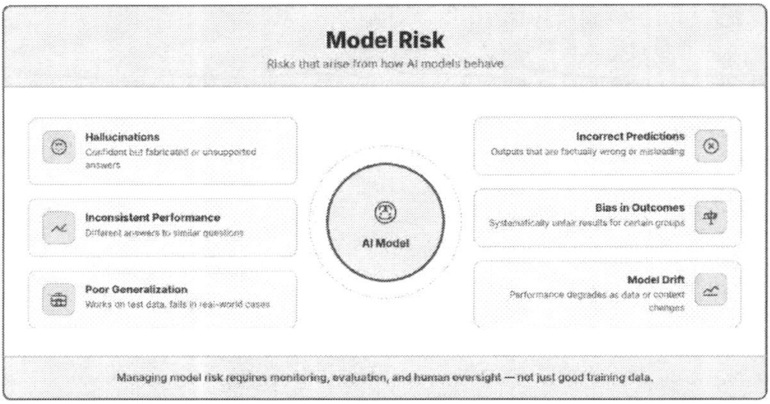

Figure 5 Model Risk

Manager's rule: Model risk is not a technical issue — it is an operational one.

2. Data Risk

Data is the fuel for AI — and a major source of risk.

Examples:

- Sensitive data exposure
- Unauthorized access
- Poor data quality
- Incomplete or biased datasets
- Data leakage
- Lack of lineage or documentation

Manager's rule: If the data is risky, the AI is risky.

---

3. Operational Risk

Operational risk arises from how AI is used in workflows.

Examples:

- Over-reliance on AI
- Misuse or misunderstanding
- Poorly designed human-in-the-loop
- Workflow bottlenecks
- Lack of training
- Inadequate monitoring

Manager's rule: AI risk increases when workflows are unclear.

---

4. Compliance & Regulatory Risk

AI must comply with laws, regulations, and policies.

Examples:

- HIPAA (healthcare)
- GDPR (privacy)
- State AI regulations

- Federal AI guidance
- Sector-specific rules
- Internal policies

Manager's rule: Compliance is not a checkbox — it is a continuous process.

Diagram: The Four Categories of AI Risk

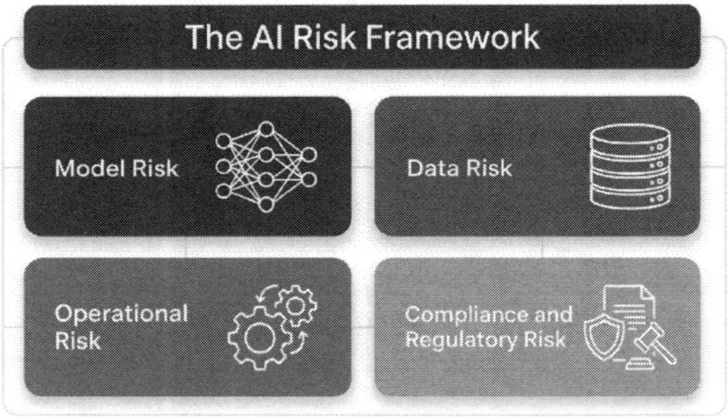

12.4 The AI Risk Mitigation Framework

To manage risk effectively, use this five-step framework:

Let's break it down.

---

1. Identify

Identify risks early — during strategy and design.

Ask:

- What could go wrong?
- Who could be harmed?
- What data is involved?
- What decisions does AI influence?

---

2. Assess

Assess the likelihood and impact.

Use a simple scale:

- Low
- Medium
- High

High-impact risks require strong safeguards.

---

3. Mitigate

Mitigation strategies include:

- Human-in-the-loop
- Guardrails
- Access controls
- Prompt constraints
- Data minimization
- Bias testing
- Workflow redesign

---

4. Monitor

Monitoring must be continuous.

Monitor:

- Drift
- Errors
- User feedback
- Performance
- Compliance

---

5. Respond

Define:

- Incident response
- Escalation paths
- Communication plans
- Remediation steps

A good response plan reduces harm.

12.5 Security Considerations for AI Systems

AI introduces new security challenges.

Prompt injection — An attacker manipulates an AI model by crafting inputs that cause it to ignore instructions, reveal sensitive information, or perform unintended actions. This is especially dangerous in agentic or workflow-integrated systems, where the model can trigger downstream actions.

Model extraction — An adversary attempts to recreate or steal a proprietary model by repeatedly querying it and inferring its parameters, training data, or decision boundaries. This can compromise intellectual property and expose sensitive patterns learned from private data.

Data poisoning — Malicious actors insert corrupted, biased, or adversarially crafted data into the training pipeline, causing the model to learn incorrect behaviors. Poisoning can degrade accuracy, introduce vulnerabilities, or cause targeted misclassifications.

Unauthorized access — Attackers gain access to model endpoints, training data, logs, or system controls. This can lead to data breaches, manipulation of model behavior, or unauthorized use of compute resources.

API misuses. Legitimate APIs are used in unintended or harmful ways, such as excessive querying, chaining calls to bypass safeguards, or exploiting poorly validated inputs. This can lead to system overload, data leakage, or unintended actions.

Supply chain vulnerabilities — Risks introduced through third-party models, datasets, libraries, or infrastructure. Compromised dependencies can propagate vulnerabilities across the entire AI stack, making supply chain security a critical part of AI governance.

Security safeguards:

Security safeguards such as access controls, encryption, input validation, rate limiting, audit logs, and zero-trust architecture form the foundation of a resilient AI environment. Each one addresses a different failure mode—preventing unauthorized entry, protecting sensitive data, ensuring only clean inputs reach the model, stopping abusive usage patterns, creating traceability, and eliminating implicit trust across systems. Together, they reduce the likelihood that an attacker can exploit AI systems or that internal errors go undetected.

Access controls — Restrict who can view, modify, or interact with AI systems using role-based permissions, least-privilege principles, and strong authentication. This prevents unauthorized users from reaching sensitive models, data, or configuration settings.

Encryption — Protects data at rest and in transit so that even if intercepted or stolen, it cannot be read or tampered with. Encryption is essential for safeguarding training data, model outputs, and API traffic.

Input validation — Ensures that only properly formatted, safe, and expected inputs reach the model or downstream systems. This reduces risks like prompt injections, malicious payloads, or malformed data that could trigger unintended behavior.

Rate limiting — Caps how frequently users or systems can call an API or model endpoint. This prevents brute-force attacks, model extraction attempts, and system overloads that could degrade performance or availability.

Audit logs — Record who accessed the system, what actions they took, and when. Logs support incident investigation, compliance reporting, and early detection of suspicious behavior.

Zero trust architecture — Removes implicit trust between systems, requiring continuous verification of identity, device health, and permissions for every request. This minimizes lateral movement and reduces the blast radius of any breach.

Security must be built into the AI lifecycle.

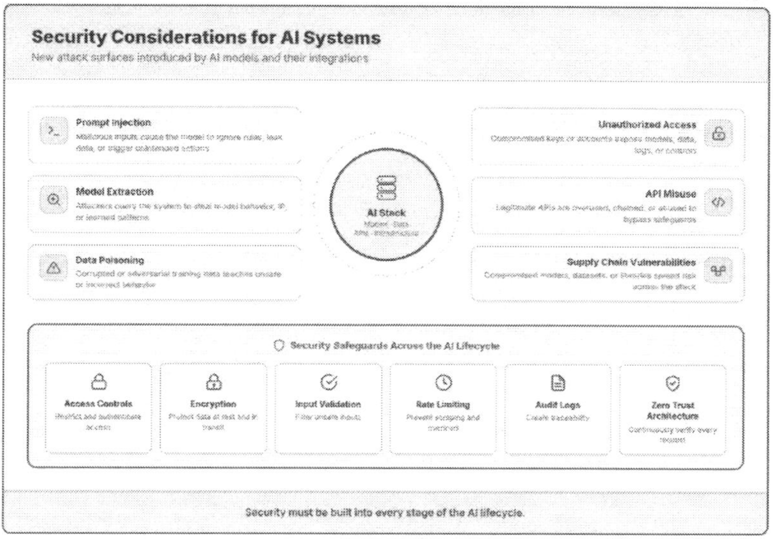

12.6 AI Compliance: What Managers Need to Know

Compliance is not just legal — it is operational.

Managers must ensure:

- Data use is lawful
- AI decisions are explainable
- Documentation is complete
- Risks are logged
- Oversight is defined
- Policies are followed

Compliance teams should be partners — not gatekeepers.

The AI Risk Register

Every organization should maintain an AI Risk Register.

It includes:

- Identified risks
- Severity
- Likelihood
- Mitigation steps
- Owners
- Status

This becomes the single source of truth for risk management.

Case Example: Healthcare

Risk: Hallucinations in clinical summaries

Mitigation: Human-in-the-loop + transparency + monitoring

Case Example: Public Sector

Risk: Bias in eligibility screening

Mitigation: Bias testing + override mechanisms + audit logs

Case Example: Enterprise

Risk: Prompt injection in customer service AI

Mitigation: Input validation + guardrails + monitoring

The Responsible AI Lens: Risk Is About People

Risk management is not about protecting the organization — it is about protecting people.

Ask:

- Who could be harmed?
- How do we prevent harm?
- How do we ensure fairness?
- How do we maintain trust?

Risk management is leadership.

12.7 Summary — Key Takeaways

- AI introduces new categories of risk.
- Risk must be managed proactively, not reactively.
- The four categories of risk are model, data, operational, and compliance.
- The five-step mitigation framework is identify → assess → mitigate → monitor → respond.
- Security must be built into the AI lifecycle.
- Compliance is continuous.
- Risk management protects people and the mission.

12.8 Workbook 12: *Risk Compliance & Security.*

This workbook helps you apply risk management principles to your use cases.

Exercise 1 — Risk Identification

List risks for your use case.

Risk Category	Risks Identified	Notes
Model		
Data		
Operational		
Compliance		

Exercise 2 — Risk Assessment

Rate each risk (Low/Medium/High).

Risk	Likelihood	Impact	Priority
1.________			
2.________			

Exercise 3 — Mitigation Plan

Risk	Mitigation Strategy	Owner	Timeline
1.________			

Risk	Mitigation Strategy	Owner	Timeline
2.________			

Exercise 4 — Monitoring Plan

Define how you will monitor risk.

Metrics:

- ______________________________
- ______________________________

Tools:

- ______________________________

Frequency:

- ______________________________

Exercise 5 — Incident Response Plan

Define your response plan.

Trigger:

____

Actions:

- ______________________________
- ______________________________

Escalation Path:

12.9 1-Page Handout: The AI Risk Mgmt Blueprint

Perfect for workshops, leadership briefings, or onboarding.

THE AI RISK MANAGEMENT BLUEPRINT

The Four Categories of Risk

- Model
- Data
- Operational
- Compliance

The Five-Step Framework

- Identify
- Assess
- Mitigate
- Monitor
- Respond

Security Safeguards

- Access controls
- Encryption
- Input validation
- Audit logs
- Zero-trust

Compliance Essentials

- Documentation

- Transparency
- Oversight
- Policy alignment

Principle

Risk management protects people, the mission, and the organization.

Part V — Leading the AI Transformation

Part V brings the book into its most human and strategic territory: the work of *leading* an AI-enabled organization. While earlier sections focused on understanding AI, designing pilots, scaling systems, and governing risk, this final part turns to the leadership behaviors, cultural shifts, and organizational capabilities required to make AI transformation stick. It explores how managers set direction, shape incentives, build trust, and guide teams through uncertainty. Readers learn how to communicate a compelling AI vision, develop talent, redesign workflows, and create the psychological safety needed for experimentation. This section emphasizes that AI transformation is ultimately a people transformation—one that succeeds only when leaders model curiosity, transparency, and responsible decision-making.

Part V also equips leaders with practical tools for sustaining momentum over the long term. It covers how to build cross-functional collaboration, manage change fatigue, and create feedback loops that keep AI initiatives aligned with mission and measurable outcomes. It highlights the importance of continuous learning, adaptive governance, and the ability to translate technical progress into operational value. By the end of this section, readers will understand that leading the AI transformation is not about mastering algorithms—it's about cultivating the systems, culture, and leadership practices that allow AI to elevate the entire organization.

13 Building AI Talent, Skills, and Org. Capability

Note that chapter 13 is pivotal because AI introduces a kind of change that traditional transformation playbooks don't address. Unlike past technology shifts, AI alters workflows, decision-making patterns, and job identities at the same time—often in ways that feel ambiguous or threatening to employees. This chapter explains why leaders need a **different change playbook** for AI, one rooted in transparency, empathy, and continuous communication rather than top-down announcements or technical training alone. It shows how to communicate AI's value in a grounded, credible way—avoiding hype while still helping teams see the practical benefits for mission, quality, and daily work. By reframing AI as a tool that augments people rather than replaces them, leaders can build trust early and reduce the anxiety that often derails adoption.

The chapter also focuses on the human side of transformation: **upskilling, workforce evolution, and managing resistance**. It provides practical guidance for helping employees build new capabilities, redesigning roles around human-AI collaboration, and creating safe spaces for questions and concerns. It addresses the emotional realities of AI adoption—fear of job loss, uncertainty about new expectations, and skepticism about automation—and equips managers with strategies to respond constructively. By the end of Chapter 13, leaders understand how to guide their teams through AI-driven change with clarity, compassion, and confidence, ensuring that people feel supported rather than disrupted as the organization moves forward.

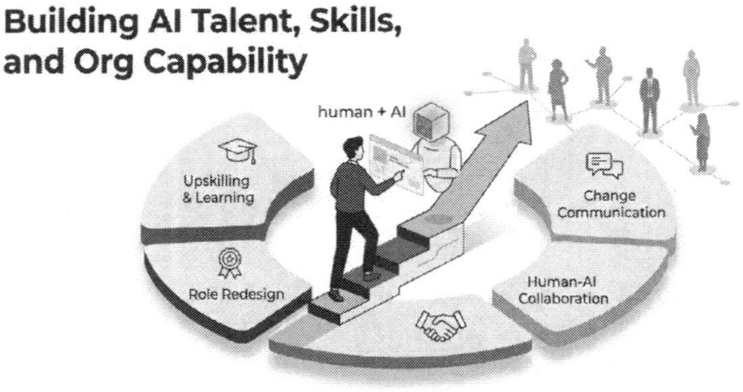

13.1 Opening Scenario: The Capability Gap

Your AI pilots are working. Your roadmap is clear. Your operating model is defined. Your governance is strong.

Then you hit a wall:

- "We don't have enough people who understand AI."
- "Our teams don't know how to use these tools."
- "We don't have the right skills to maintain the models."
- "We're relying too heavily on vendors."
- "We don't have internal champions."
- "We don't know how to train everyone."

This is the **capability gap** — the moment when organizations realize that AI is not just a technology transformation.

It is a **skills transformation**, a **workforce transformation**, and a **culture transformation**.

This chapter provides the blueprint for building the talent and organizational capabilities required to make AI sustainable.

13.2 Why AI Capability Matters Most Competitively.

Technology changes fast. Skills endure.

Organizations that succeed with AI:

- Invest in people.
- Build internal capability.
- Develop AI-literate leaders.
- Create cross-functional teams.
- Train the workforce.
- Build reusable knowledge.
- Reduce vendor dependency.

Organizations that fail:

- Treat AI as a technical project.
- Outsource everything.
- Underinvest in training.
- Ignore change management.
- Fail to build internal champions.

AI capability is not optional — It is strategic.

13.3 The Three Layers of AI Capability

Every organization needs capability at three levels:

1. AI Literacy (Everyone)
2. AI Practitioners (Specialists)
3. AI Leaders (Decision-Makers)

Let's break them down.

---

1. AI Literacy (Everyone)

AI literacy is the foundation.

Every employee should understand:

- What AI can and cannot do
- How to use AI tools safely
- How to evaluate AI outputs
- How to escalate concerns
- How AI affects their workflow
- How to maintain human oversight

AI literacy is not technical — it is practical.

---

2. AI Practitioners (Specialists)

These are the people who build, integrate, and maintain AI systems.

Roles include:

- Data scientists
- Machine Learning Engineers
- Prompt engineers

- AI product managers
- Data engineers
- Responsible AI specialists
- Workflow designers

Practitioners turn strategy into reality.

---

3. AI Leaders (Decision-Makers)

AI leaders make strategic decisions about:

- Use case prioritization.
- Investment
- Governance
- Risk
- Scaling
- Operating model design

AI leaders do not need to be technical — but they must be AI-fluent

---

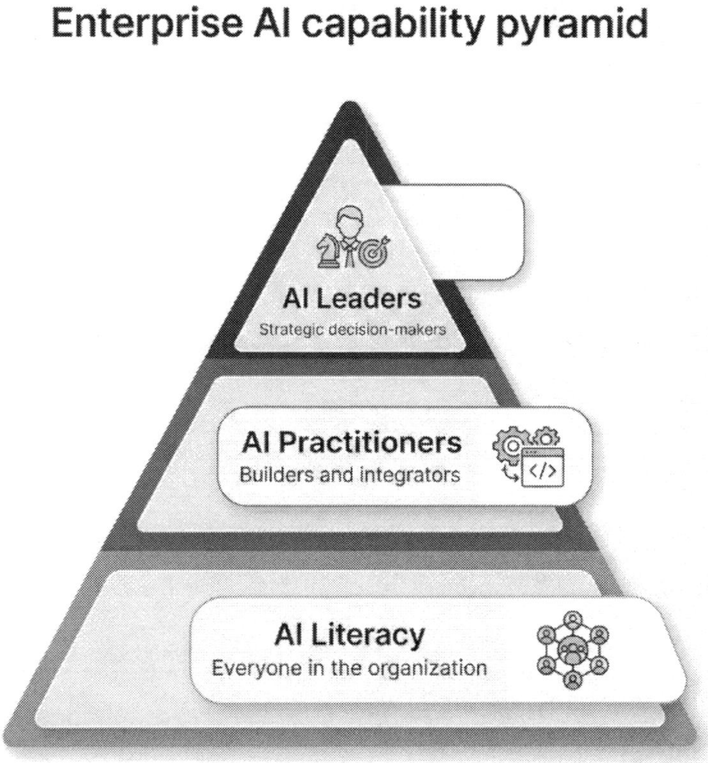

Figure 6 The Three Layers of AI Capability

13.4 The AI Capability Maturity Model

Organizations evolve through four stages of AI capability:

Stage 1 — Awareness

Stage 2 — Experimentation

Stage 3 — Operationalization

Stage 4 — Institutionalization

Let's break them down.

Stage 1 — Awareness

Characteristics:

- Limited understanding of AI
- Ad hoc experimentation
- No governance
- No training
- Goal: Build literacy and alignment.

Stage 2 — Experimentation

Characteristics:

- Pilots underway
- Early wins
- Emerging governance
- Growing interest

Goal: Build practitioner capability.

Stage 3 — Operationalization

Characteristics:

- AI embedded in workflows
- Governance functioning
- Clear operating model

• Dedicated teams

Goal: Build leadership capability.

Stage 4 — Institutionalization

Characteristics:

- AI scaled across the enterprise
- Mature governance
- Strong internal talent
- Continuous improvement

Goal: Sustain long-term capability.

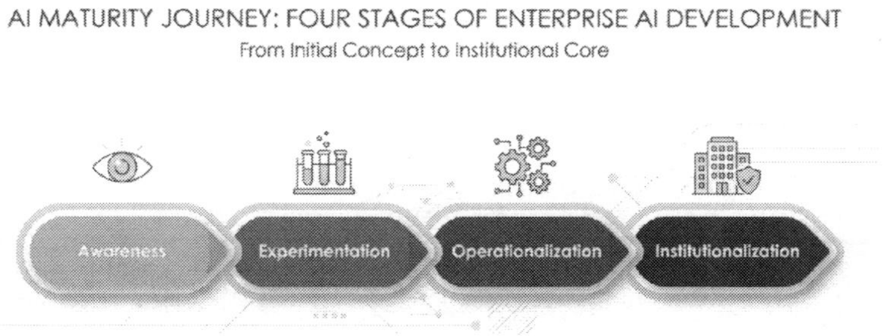

Figure 7 AI Capability Maturity Model

13.5 Building Your AI Talent Strategy

Your talent strategy should include:

1. Skills Assessment

Identify current skills and gaps.

2. Training Programs

Develop role-specific training.

3. Hiring Strategy

Fill critical gaps with external talent.

4. Internal Mobility

Upskill existing employees.

5. Vendor Strategy

Use vendors strategically — not dependently.

6. Communities of Practice

Create cross-functional learning groups.

7. Champions Network

Empower early adopters to lead change.

13.6 The AI Skills Matrix

Use this matrix to define required skills.

Role	Skills Needed
AI Leaders	Strategy, governance, risk, prioritization
Practitioners	Modeling, prompting, data engineering, evaluation
Workflow Owners	Process design, oversight, and adoption
End Users	AI literacy, tool usage, escalation

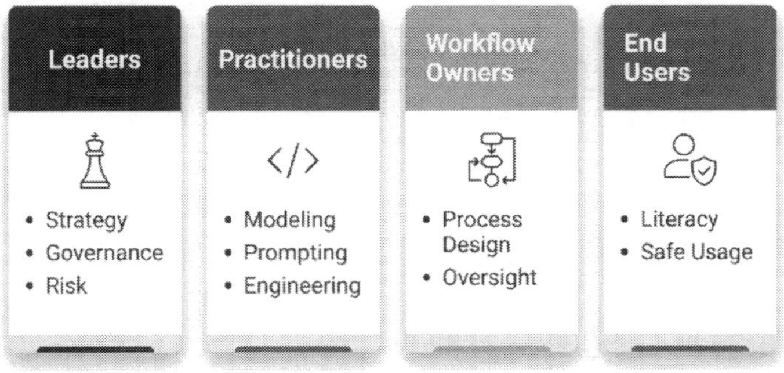

Figure 8 The AI Skills Matrix

13.7 Training Programs That Work

Effective AI training is:

- Hands-on
- Workflow-specific
- Role-based
- Scenario-driven
- Continuous

Training should include:

- AI fundamentals
- Responsible AI
- Workflow integration
- Tool usage
- Escalation procedures
- Case studies

13.8 Building a Champions Network

Champions accelerate adoption.

Champions should:

- Be early adopters
- Understand workflows
- Support peers
- Provide feedback
- Identify risks
- Celebrating wins

Champions turn AI from a project into a movement.

13.9 Case Example: Healthcare

Challenge: Clinicians are overwhelmed by documentation

- **Solution**: AI literacy + champions + workflow training
- **Result:** 30% reduction in documentation time

Case Example: Public Sector

Challenge: Low adoption of AI tools

- **Solution:** Role-based training + champions network
- **Result:** 55% improvement in adoption

Case Example: Enterprise

Challenge: Over-reliance on vendors

- **Solution:** Build an internal practitioner team
- **Result:** Faster scaling + lower cost

The Responsible AI Lens: Capability Protects People

Capability ensures:

- AI is used safely
- Oversight is strong
- Risks are identified early
- Workflows remain human-centered
- Decisions remain accountable

Capability is responsibility.

13.10 Summary — Key Takeaways

- AI capability is the real competitive advantage.
- Every organization needs literacy, practitioners, and leaders.
- Capability evolves through four maturity stages.
- Training must be role-based and workflow-specific.
- Champions accelerate adoption.
- Capability protects people and the mission.

13.11 Workbook 13:Build AI Talent & Capability

This workbook helps you build your talent and capability strategy.

Exercise 1 — Capability Assessment

Rate your organization (1–5).

Capability Area	Score	Notes
AI Literacy		
Practitioner Skills		
Leadership Fluency		
Training Programs		
Governance Skills		
Change Management		

Exercise 2 — Skills Gap Analysis

List gaps.

Role	Current Skills	Gaps	Priority
Leaders			
Practitioners			
Workflow Owners			
End Users			

Exercise 3 — Training Plan

Define training needs.

Role	Training Needed	Format	Owner
Leaders			
Practitioners			
End Users			

Exercise 4 — Champions Network

Identify champions.

Name	Team	Strengths	Role
1.__________			
2.__________			

Exercise 5 — Capability Roadmap

Define your 6-month capability plan.

Area	Actions	Owner	Timeline
Training			
Hiring			
Governance			
Champions			

---

14 Driving Adoption: Change Management for AI

Note that chapter 14 stands out as a pivotal moment, moving attention away from single AI ventures and towards the cultural bedrock that shapes organizational success with AI. While technology is essential, true transformation hinges on people—their routines, attitudes, and everyday actions determine whether AI becomes a reliable asset or fails to advance. This chapter explores the **habits of AI-mature organizations, showing how they normalize experimentation, encourage data-driven decision-making, and treat learning as a continuous practice rather than a one-time** training event. It highlights why culture is the true multiplier: when teams are curious, open to change, and empowered to explore AI safely, adoption accelerates naturally and sustainably.

Equally important, Chapter 14 explains how to **create AI champions**, design incentives that reinforce adoption, and measure cultural progress in ways leaders can act on. It shows how to identify early adopters, equip them with the right tools, and position them as peer-level influencers who help others build confidence. It also examines the role of incentives—recognition, time savings, career growth, and mission alignment—in shaping behavior. Finally, it provides practical ways to assess cultural readiness and track shifts over time, ensuring leaders can see where momentum is building and where additional support is needed. Together, these insights help organizations build a culture where AI is not just accepted but actively embraced.

14.1 Opening Scenario: The Adoption Cliff

Your AI solution is ready. The pilot succeeded. The metrics look great. Leadership is excited.

Then you deploy it to the broader organization — and adoption collapses.

- People don't trust the AI
- They revert to old workflows
- They don't understand how to use the tool
- They fear job loss
- They don't know when to rely on AI
- They don't know how to escalate issues
- They don't feel included in the process

This is the **adoption cliff** — where AI initiatives fail not because of technology, but because of people.

AI adoption is not automatic. It must be **designed**, **supported**, and **led**.

This chapter gives you the blueprint to drive adoption across your organization.

14.2 Why Adoption Is the Hardest Part of AI

AI changes:

- How people work
- How decisions are made
- How workflows operate
- How teams collaborate
- How risk is managed
- How performance is measured

People resist change when:

- They don't understand it
- They don't trust it
- They don't feel prepared
- They don't feel safe
- They don't feel included

Adoption is not a training problem; it is a Human Problem.

14.3 The AI Adoption Curve

Every organization follows the same adoption curve:

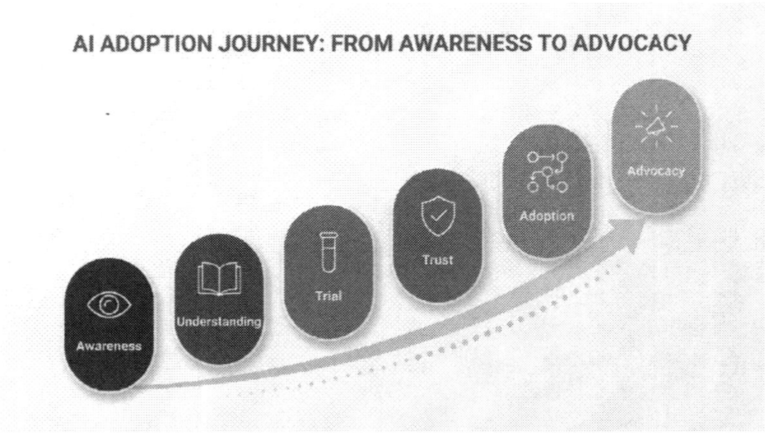

Let's break it down.

---

1. Awareness

People know AI is coming — but don't know what it means.

Goal: Reduce fear.

---

2. Understanding

People understand what AI will do — and what it won't.

Goal: Build clarity.

---

3. Trial

People try AI in a safe, low-risk environment.

Goal: Build familiarity.

---

4. Trust

People see that AI is reliable and safe.

Goal: Build confidence.

---

5. Adoption

People use AI consistently in their workflows.

Goal: Build habits.

---

6. Advocacy

People champion AI and help others adopt it.

Goal: Build momentum.

14.4 The Five Levers of AI Change Management

To drive adoption, you must activate five levers:

1. Communication
2. Training
3. Workflow Integration
4. Support & Feedback Loops
5. Leadership & Culture

Let's break them down.

---

1. Communication

Communication must be:

- Clear

- Frequent
- Transparent
- Human-centered
- Mission-aligned

Communicate:

- Why AI is being adopted
- What problems does it solve
- What will it change
- What it will NOT change
- How people will be supported

Silence creates fear. Communication creates trust.

---

2. Training

Training must be:

- Role-based
- Workflow-specific
- Hands-on
- Scenario-driven
- Continuous

Training should include:

- How to use the AI tool
- How to evaluate AI outputs
- How to escalate issues
- How to maintain oversight
- How to integrate AI into daily work

Training is not a one-time event.

It is a continuous process.

---

3. Workflow Integration

AI adoption fails when:

- The workflow is unclear
- The workflow is inconsistent
- AI is bolted on instead of integrated
- People don't know when to use AI
- People don't know when NOT to use AI

Workflow integration must be intentional.

---

4. Support & Feedback Loops

People need:

- Help when they get stuck
- A place to ask questions
- A way to report issues
- A way to suggest improvements
- A way to feel heard

Support channels include:

- Office hours
- Champions network
- Help desk
- Slack/Teams channels
- Feedback forms

Feedback loops accelerate improvement.

5. Leadership & Culture

Leaders must:

- Model AI usage
- Reinforce expectations
- Celebrating wins
- Address fears
- Provide psychological safety

Culture determines whether AI is embraced or rejected.

14.5 The AI Adoption Playbook

Here is the playbook you will use across your organization:

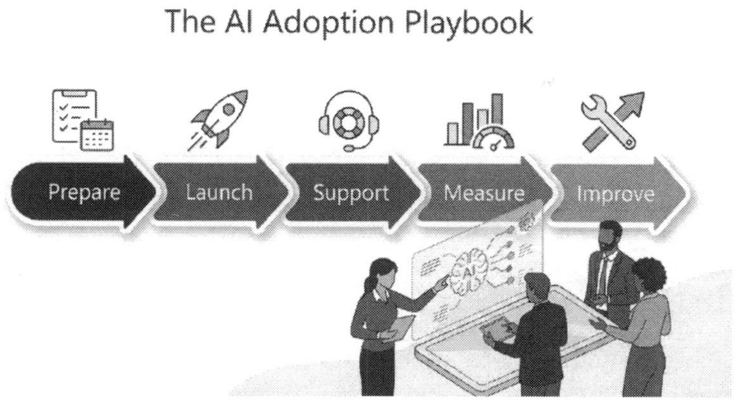

Let's break it down.

---

1. Prepare

- Identify stakeholders
- Map workflows
- Build training
- Create a communication plan
- Identify champions

---

2. Launch

- Deliver training
- Communicate expectations
- Provide quick-start guides
- Run office hours
- Monitor early usage

---

3. Support

- Provide ongoing help
- Collect feedback
- Address issues
- Reinforce workflows

---

4. Measure

Track:

- Adoption rates
- Usage patterns
- User satisfaction
- Workflow impact
- Risk incidents

---

5. Improve

- Refine workflows
- Update training
- Improvement of prompts
- Strengthening safeguards
- Celebrating wins

14.6 Case Examples

Case Example: Healthcare

Challenge: Clinicians are hesitant to trust AI

- **Solution:** Role-based training + champions + transparency
- **Result:** 30% increase in adoption

Case Example: Public Sector

Challenge: Low usage of the AI routing tool

- **Solution:** Workflow redesign + communication + support
- **Result:** 55% improvement in adoption

Case Example: Enterprise

Challenge: Over-reliance on old processes

- **Solution:** Leadership modeling + incentives
- **Result:** Rapid adoption across teams

The Responsible AI Lens: Adoption Is Safety

Adoption is not just about usage — it is about safety.

When people:

- Understand AI
- Trust AI
- Know how to use AI
- Know how to escalate issues
- Know how to maintain oversight

AI becomes safer, more reliable, and more aligned with
Mission Outcomes.

14.7 Summary — Key Takeaways

- Adoption is the hardest part of AI — and the most important.

- The AI Adoption Curve moves from awareness to advocacy.
- The five levers of adoption are communication, training, workflow integration, support, and leadership.
- Adoption requires a structured playbook.
- Adoption is a safety mechanism — not just a change management task.

Workbook 14 — Driving AI Adoption

This workbook helps you design your adoption strategy.

Exercise 1 — Adoption Readiness Assessment

Rate your organization (1–5).

Area	Score	Notes
Communication		
Training		
Workflow Integration		
Support		
Leadership		
Culture		

Exercise 2 — Stakeholder Mapping

Identify key stakeholders.

Stakeholder	Role	Influence	Needs
1.__________			
2.__________			

Exercise 3 — Training Plan

Define training needs.

Role	Training Needed	Format	Owner
End Users			
Workflow Owners			
Leaders			

Exercise 4 — Communication Plan

Define your communication strategy.

Key Messages:

- ____________________
- ____________________

Channels:

• ____________________

Cadence:

• ____________________

Exercise 5 — Support & Feedback Plan

Define support channels.

Channel	Purpose	Owner
Office Hours		
Help Desk		
Slack/Teams		

Exercise 6 — Adoption Metrics

Define how you will measure adoption.

Metrics:

• ____________________

• ____________________

• ____________________

One-Page Handout: The AI Adoption Blueprint

Perfect for workshops, leadership briefings, or onboarding.

14.8 The AI Adoption Blueprint

The Adoption Curve

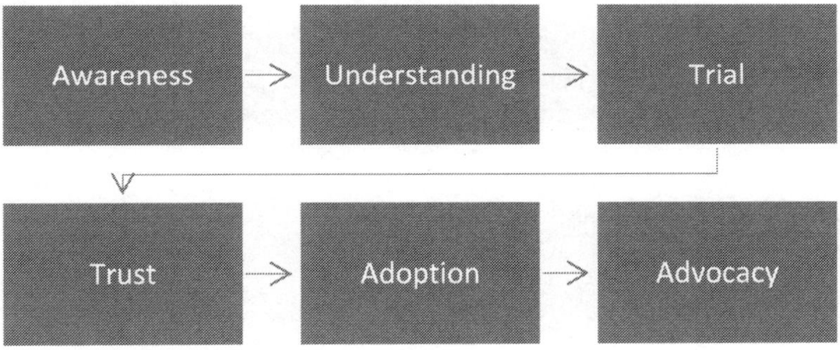

Five Levers

- Communication
- Training
- Workflow Integration
- Support & Feedback
- Leadership & Culture

Adoption Playbook

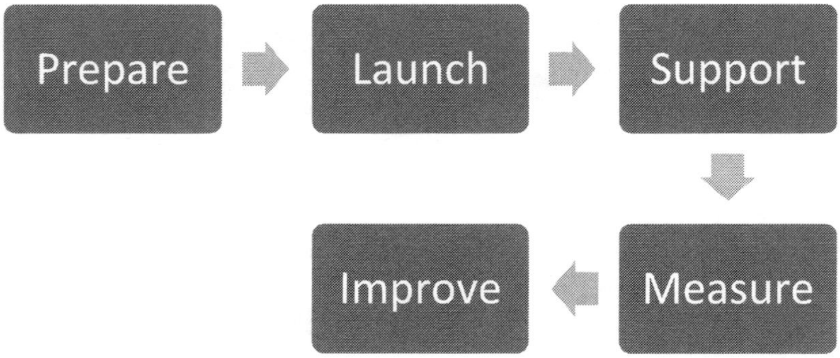

Principle

Adoption is not automatic — it must be designed, supported, and led.

15 The AI Manager's Playbook

AI transformation succeeds when managers adopt a consistent rhythm of leadership—weekly habits that keep teams aligned, monthly reviews that surface risks early, and quarterly checkpoints that ensure the organization is learning, adapting, and scaling responsibly. Chapter 15 distills these rhythms into a practical playbook: a set of repeatable rituals, communication patterns, and partnership models that help managers lead AI initiatives with clarity and confidence. This chapter is designed to be used, not just read. It gives leaders a cadence they can rely on, even as technology evolves.

Just as importantly, this chapter shows managers how to partner effectively with technical teams, run structured AI reviews, and stay ahead of the curve without getting lost in hype. AI leadership is not about mastering every model or algorithm—it's about creating the conditions where teams can experiment safely, learn quickly, and deliver measurable value. The playbook provides the scaffolding for that environment: predictable routines, shared language, and a disciplined approach to decision-making. With these tools, managers can guide their teams through uncertainty and build AI capabilities that grow stronger over time.

The AI Manager's Playbook
Visualize leadership Thythms and Collaboration
Weekly Team Rituals
Monthly Risk & Learning Reviews
Quarterly Strategy Checkpoints
Partnership with Technical Teams

15.1 The AI Manager's Operating Rhythm

Weekly Rituals

Weekly rituals keep momentum high and ensure that AI work stays connected to real operational needs.

- **Weekly AI Standup (15 minutes)**

- What did we learn this week?
- What changed in the workflow?
- Any new risks or unexpected behaviors?
- What's the next small experiment?

- **Workflow Check-ins**

- Confirm that the workflow being automated hasn't changed.
- Validate that human-in-the-loop checkpoints are functioning.
- Capture frontline feedback.

- **Data Quality Pulse**

- Spot anomalies early.
- Flag missing or drifting data.
- Identify new data sources that could improve performance.

Monthly Rituals

Monthly rituals help managers zoom out and evaluate progress, risks, and alignment.

- **Monthly AI Review**

- Performance vs. success metrics
- Error patterns and edge cases
- User adoption and satisfaction

- Compliance and safety checks
- Roadmap adjustments
- **Cross-Functional Sync**
- Align with IT, security, compliance, and operations.
- Review shared dependencies and risks.
- Identify opportunities for reuse.

Quarterly Rituals

Quarterly rituals ensure strategic alignment and long-term sustainability.

- **Quarterly AI Strategy Review**
- Revisit mission alignment.
- Evaluate scaling opportunities
- Assess governance maturity
- Refresh the 90-day roadmap
- **Capability Building Review**
- Skills gained
- Skills needed
- Training and upskilling plans
- Hiring or vendor support needs

15.2 How to Run an AI Review

AI reviews are the backbone of responsible, measurable progress. They ensure that AI systems deliver value, operate safely, and evolve with the organization.

15.3 The AI Review Agenda

AI REVIEW AGENDA

---

1. Mission Alignment Check

Is the use case still tied to a real problem?

2. Performance Review

Metrics, accuracy, error rates, drift signals

3. Workflow Health

Has the workflow changed?

Are humans-in-the-loop engaged?

4. Risk & Safety Review

Bias checks

Security posture

Compliance updates

5. Adoption & Feedback

User satisfaction

Friction points

Training needs

6. Roadmap Decisions

Continue, pause, scale, or retire

15.4 The Four Outcomes of an AI Review

- **Continue** — The system is performing well and aligned with the mission.
- **Scale** — The system is ready for broader deployment or reuse.
- **Pause** — Issues require investigation before proceeding.
- **Retire** — The system no longer solves a meaningful problem.

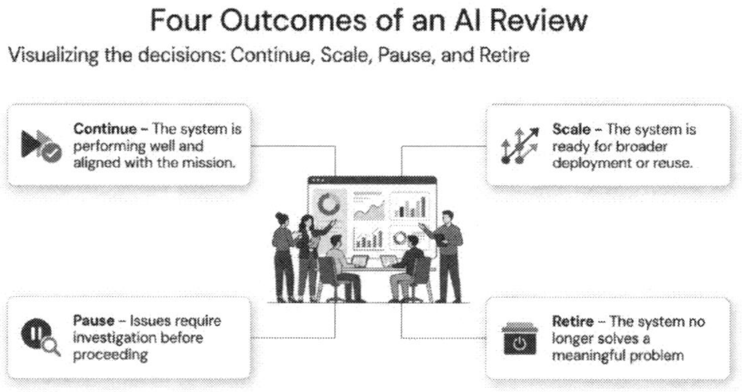

15.5 How to Partner with Technical Teams

Strong AI leadership depends on strong partnerships. Managers don't need to be technical experts—they need to create clarity, remove obstacles, and ensure alignment.

The Manager's Role

- Define the problem, not the solution.
- Provide workflow context and operational constraints.
- Ensure access to subject-matter experts.
- Translate business needs into measurable outcomes.

- Protect teams from shifting priorities.

The Technical Team's Role

- Assess feasibility and risks.
- Build and evaluate models.
- Integrate systems into workflows.
- Monitor performance and drift.
- Implement safeguards and controls.

15.6 The Collaboration Model

COLLABORATION LOOP

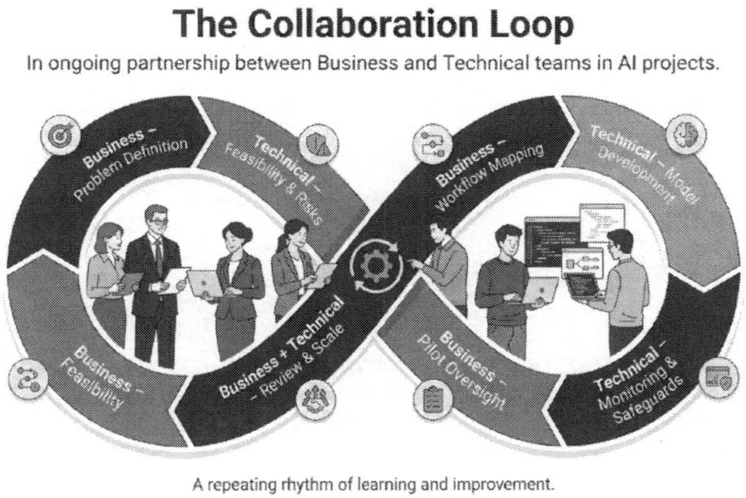

Business → Problem Definition

Technical → Feasibility & Risks

Business → Workflow Mapping

Technical → Model Development

Business → Pilot Oversight

Technical → Monitoring & Safeguards

Business + Technical → Review & Scale

The loop repeats with every iteration, creating a shared rhythm of learning and improvement.

15.7 How to Stay Ahead of the Curve

AI evolves quickly, but managers don't need to chase every trend. They need a disciplined approach to staying informed and making strategic decisions.

The Three Horizons of AI Awareness

- **Horizon 1: Operational Awareness**
    - What affects your current systems?
    - What new risks or safeguards matter?
- **Horizon 2: Strategic Awareness**
    - What capabilities could unlock new workflows?
    - What vendor or platform changes are emerging?
- **Horizon 3: Exploratory Awareness**
    - What's experimental but worth watching?
    - What could reshape your industry in 2–5 years?

The Manager's Learning Cadence

- 15 minutes weekly: scan trusted sources
- 1 hour monthly: deep dive into one topic
- Quarterly: attend a workshop, briefing, or training

- Annually: refresh the AI strategy with new insights

15.8 Workbook 15: The AI Manager's Playbook

Use this workbook to turn the chapter into action.

Exercise 1: Define Your AI Leadership Rhythm

Weekly Rituals I will adopt:

1.

2.

3.

Monthly Rituals I will adopt:

1.

2.

3.

Quarterly Rituals I will adopt:

1.

2.

3.

Exercise 2: Build Your AI Review Template

Success metrics to track:

-

-

Risks to monitor:

-
-

Workflow checkpoints:

-
-

Exercise 3: Map Your Technical Partnerships

Key technical partners:

-
-

What they need from me:

-
-

What I need from them:

-
-

Exercise 4: Staying Ahead of the Curve

Sources I will follow:

-
-

Topics I need to learn next:

-
-

Part VI – Tools, Templates, and Case Studies

Part□VI anchors the book in practical execution by giving managers the **tools, templates, and real-world case studies** needed to turn strategy into action. While earlier sections build understanding, alignment, and decision-making capability, this part closes the gap between knowing and doing by providing ready-to-use assessments, scorecards, roadmaps, and pilot design frameworks that accelerate implementation and reduce organizational friction. The case studies then translate these tools into lived examples, showing how different sectors navigated constraints, risks, and trade-offs to achieve measurable outcomes. Together, these resources form a repeatable operating system for AI adoption—one that equips leaders not just to plan responsibly, but to execute confidently, avoid common pitfalls, and scale AI in a way that is sustainable, safe, and aligned with their mission.

16 Templates and Frameworks

Part of becoming a serious AI manager is having a set of tools you can rely on—templates that reduce ambiguity, frameworks that create shared language, and structured canvases that help teams move from ideas to execution. Chapter 16 brings together the most important tools introduced throughout the book and presents them in a format you can use immediately. These templates are designed to accelerate decision-making, improve alignment, and ensure that every AI initiative is grounded in mission, value, feasibility, and responsible practice.

This chapter is intentionally practical. Each template is built for real-world use: in workshops, planning sessions, executive briefings, and day-to-day management. Whether you're assessing readiness, scoring use cases, designing pilots, or building a 90-day roadmap, these frameworks give you a repeatable structure that scales with your organization. They also help you build a culture of clarity—where teams know what "good" looks like and can move quickly without sacrificing safety or rigor.

16.1 AI Readiness Assessment Framework

This framework helps leaders evaluate whether the organization is prepared to adopt AI responsibly and effectively. It covers four dimensions: **data**, **people**, **processes**, and **culture**.

AI Readiness Assessment (Score 1–5 per dimension)

AI READINESS ASSESSMENT

DATA

1 = Fragmented, inaccessible

3 = Partially structured, inconsistent quality

5 = Reliable, governed, accessible, documented

PEOPLE

1 = Low AI literacy, unclear roles

3 = Some champions, uneven skills

5 = Skilled teams, clear ownership, ongoing training

PROCESSES

1 = Manual, variable, undocumented

3 = Some standardization, some automation

5 = Stable, documented, automation-ready workflows

CULTURE

1 = Fearful, resistant, low trust

3 = Mixed attitudes, pockets of support

5 = Curious, experimental, psychologically safe

TOTAL READINESS SCORE: ___ / 20

Interpretation

- **16–20** → Ready for pilots and early scaling
- **10–15** → Ready for targeted pilots with support
- **<10** → Focus on foundational readiness before pilots

16.2 Use Case Scorecard

This scorecard helps teams evaluate and prioritize AI opportunities based on value, feasibility, risk, and readiness.

Use Case Scorecard (1–5 scale)

USE CASE SCORECARD

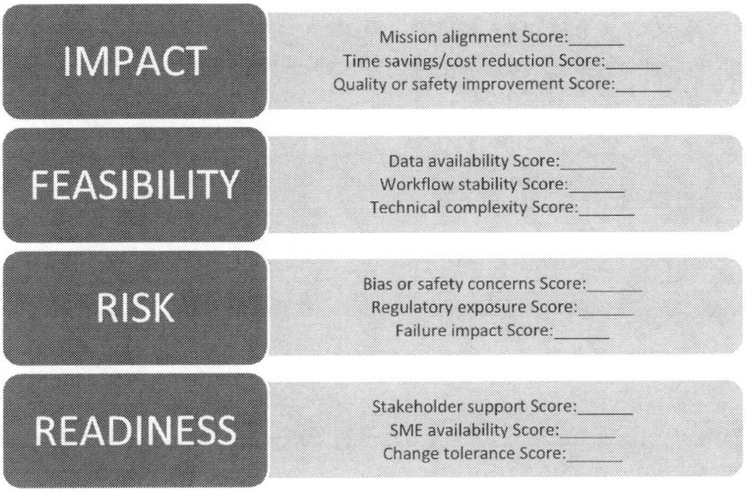

TOTAL SCORE: ___ / 20

16.3 Prioritization Guidance

- **16–20** → High-priority candidate
- **12–15** → Medium priority; refine scope
- **<12** → Low priority or revisit later

16.4 90-Day Roadmap Template

This template structures the first 90 days of an AI initiative into three horizons: **Foundation**, **Pilot**, and **Scale Preparation**.

90-Day AI Roadmap

90-DAY AI ROADMAP

HORIZON 1 (Days 1–30): FOUNDATION

• Confirm mission alignment

• Map workflow and data sources

• Establish success metrics

• Validate feasibility and risks

• Secure SMEs and stakeholders

HORIZON 2 (Days 31–60): PILOT

• Build and test prototype

• Validate with real users

• Monitor errors and drift

• Document edge cases

• Adjust workflow and safeguards

HORIZON 3 (Days 61–90): SCALE PREP

• Evaluate pilot results

• Strengthen governance

• Build reusable components

• Prepare training and change plan

• Decide: scale, refine, or retire

16.5 Pilot Design Canvas

A one-page canvas that helps teams design AI pilots with clarity and alignment.

Pilot Design Canvas

PILOT DESIGN CANVAS

PROBLEM STATEMENT

• What problem are we solving?

WORKFLOW

• What steps are involved?

• Where does AI fit?

DATA

• What data is required?

• Is it accessible and reliable?

SUCCESS METRICS

• What does "good" look like?

RISKS & SAFEGUARDS

• What could go wrong?

• How will we mitigate?

STAKEHOLDERS

• Who owns the workflow?

• Who needs to be involved?

TIMELINE

• What are the milestones?

16.6 Responsible AI Checklist

A practical checklist to ensure every AI system is safe, fair, and aligned with organizational values.

Responsible AI Checklist

RESPONSIBLE AI CHECKLIST

FAIRNESS

☐ Bias assessment completed

☐ Representative data validated

RELIABILITY & SAFETY

☐ Stress-tested for edge cases

☐ Human-in-the-loop defined

PRIVACY & SECURITY

☐ Sensitive data protected

☐ Access controls enforced

TRANSPARENCY

☐ Model behavior documented

☐ User guidance provided

ACCOUNTABILITY

☐ Clear ownership assigned

☐ Decision logs maintained

HUMAN OVERSIGHT

☐ Escalation paths defined

☐ Monitoring plan in place

16.7 Workbook 16: Templates and Frameworks

Use this workbook to apply the templates to your own organization.

Exercise 1: Assess Your Readiness

Score each dimension (1–5) and reflect on gaps.

- Data: ___
- People: ___
- Processes: ___
- Culture: ___

What gaps must be addressed before pilots?

-
-

Exercise 2: Score Three Use Cases

Use the scorecard to evaluate three candidate use cases.

Use Case 1:	
Impact:	___
Feasibility:	___
Risk:	___
Readiness:	___
Total: ___	

Use Case 2:

Impact: ___
Feasibility: ___
Risk: ___
Readiness: ___
Total: ___

Use Case 3:

Impact: ___
Feasibility: ___
Risk: ___
Readiness: ___
Total: ___

Exercise 3: Build Your 90-Day Roadmap

List the key activities for each horizon.

Foundation (Days 1–30):

-
-

Pilot (Days 31–60):

-
-

Scale Prep (Days 61–90):

-
-

Exercise 4: Complete the Pilot Design Canvas

Fill in each section for your chosen use case.

- Problem statement:
- Workflow:
- Data:
- Success metrics:
- Risks & safeguards:
- Stakeholders:
- Timeline:

Exercise 5: Responsible AI Review

Check off each item and note any gaps.

- Fairness gaps:
- Safety gaps:
- Privacy gaps:
- Transparency gaps:
- Accountability gaps:
- Oversight gaps:

17 Case Studies

Case studies are where strategy becomes real. They show how organizations across sectors—healthcare, public sector, and enterprise—translate AI concepts into operational results. They reveal the messy middle: the unexpected challenges, cultural barriers, technical surprises, and leadership decisions that determine whether an AI initiative succeeds or stalls. This chapter presents three end-to-end examples that illustrate how serious AI managers apply the frameworks from earlier chapters to real workflows, real constraints, and real people.

Each case study highlights what worked, what failed, and why. They demonstrate that AI success is rarely about the model alone—it's about readiness, workflow clarity, governance, and leadership. These stories give you a practical lens for evaluating your own organization's strengths and blind spots, and they offer patterns you can reuse as you build your own AI roadmap.

17.1 Healthcare Case Study — Clinical + Admin. AI

A large regional health system sought to improve both clinical decision support and administrative efficiency. They launched two pilots: one for **clinical documentation summarization** and one for **prior authorization automation**.

What They Did

- Mapped the clinical documentation workflow end-to-end

- Identified high-friction administrative tasks
- Built a 90-day roadmap for each pilot
- Embedded clinicians and billing specialists in the design process
- Implemented human-in-the-loop review for all outputs

What Worked

- **Clinician trust increased** because the system summarized notes rather than generating new content.
- **Administrative time dropped by 40%** for prior authorization prep.
- **Data quality improved** because clinicians corrected summaries, feeding better data back into the system.

What Failed

- The first model struggled with **specialty-specific terminology**, requiring retraining.
- Workflow changes were not communicated early enough, causing initial resistance.
- IT underestimated integration complexity with the EHR.

Why

The clinical pilot succeeded because it respected clinician workflows and preserved human judgment. The administrative pilot succeeded because it targeted a stable, rules-based workflow. Failures stemmed from insufficient early communication and underestimating the complexity of integration—common pitfalls in healthcare AI.

17.2 Public Sector: Prg. Integrity + Citizen Services

A state agency wanted to improve **program integrity** (fraud detection) and **citizen service responsiveness** (automated inquiry triage).

What They Did

- Conducted an AI readiness assessment
- Built a use-case scorecard to prioritize opportunities
- Designed a pilot for fraud pattern detection
- Built a separate pilot for citizen inquiry classification
- Established strict governance due to regulatory constraints

What Worked

- Fraud detection accuracy improved by 22%, reducing false positives.
- Citizen inquiries were triaged in under 30 seconds, improving service levels.
- Cross-agency collaboration improved because the AI review process created shared visibility.

What Failed

- Early models surfaced biased patterns tied to demographic data, requiring immediate mitigation.
- Legacy systems slowed deployment.
- Staff feared automation would replace roles, delaying adoption.

Why

Success came from strong governance, clear mission alignment, and transparent communication. Failures stemmed from legacy infrastructure and insufficient early attention to fairness risks—both common in public-sector environments.

17.3 Enterprise Case Study: Oper. + Cust. Experience

A global logistics company aimed to improve **route optimization** and **customer support automation**.

What They Did

- Used the pilot design canvas to define scope
- Built a 90-day roadmap for each initiative
- Integrated AI into dispatch workflows
- Added a human-in-the-loop escalation path for customer support
- Conducted monthly AI reviews

What Worked

- Route optimization reduced fuel costs by **12%**.
- Customer support resolution time improved by **35%**.
- Reusable components (data pipelines, monitoring dashboards) accelerated future projects.

What Failed

- The first customer support model hallucinated responses under ambiguous prompts.
- Dispatchers resisted early versions due to a lack of transparency.
- Scaling to new regions required retraining due to different traffic patterns.

Why

The operations pilot succeeded because the workflow was stable and measurable. The customer support pilot succeeded after transparency improvements and the addition of escalation paths. Failures stemmed from insufficient prompt guardrails and regional variability—issues that required iteration, not abandonment.

17.4 Cross-Case Diag: What Worked vs. What Failed

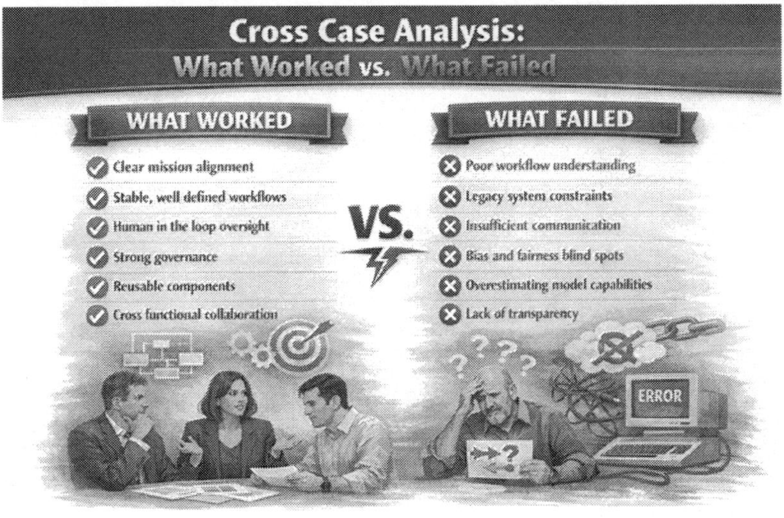

17.5 Workbook 17: Case Studies

Use this workbook to analyze your own organization through the lens of the case studies.

Exercise 1: Identify Your Sector's Patterns

Which case study most resembles your environment?

- Healthcare
- Public sector
- Enterprise

Why?

-
-

Exercise 2: What Would Work in Your Organization?

List three strengths your organization has that match the "What Worked" patterns.

1.
2.
3.

Exercise 3: What Might Fail?

List three risks your organization faces that match the "What Failed" patterns.

1.

2.

3.

Exercise 4: Build Your Case Study

Draft your own mini case study using the structure below.

- Problem:
- Workflow:
- Pilot design:
- What worked:
- What failed:
- Why:

17.6 Handout — Case Studies (One-Page)

CASE STUDIES — QUICK REFERENCE

HEALTHCARE

- Summarization + prior auth automation
- Worked: trust, time savings, data quality
- Failed: terminology, communication, integration

PUBLIC SECTOR

- Fraud detection + inquiry triage
- Worked: accuracy, service speed, governance
- Failed: bias, legacy systems, fear

ENTERPRISE

- Route optimization + support automation
- Worked: cost savings, faster resolution, reuse
- Failed: hallucinations, resistance, retraining

CROSS-CASE LESSONS

- Success = alignment, governance, workflow clarity
- Failure = bias, poor communication, legacy constraints

---

18 Closing Words

The final chapter brings the journey full circle by reminding leaders what this book has ultimately been about: building an organization that can harness AI with clarity, confidence, and purpose. Across thirty chapters, you've moved from foundational literacy to use-case selection, from readiness assessments to pilot design, from 90-day roadmaps to enterprise-scale architectures, and finally to responsible, safe, and sustainable AI. This closing chapter ties those threads together, emphasizing that AI strategy is not a one-time initiative but a continuous capability—one that strengthens as teams learn, reuse components, refine governance, and build trust across the enterprise. It reinforces that the real differentiator is not the models you deploy, but the systems, habits, and leadership practices you cultivate around them.

This chapter also looks ahead, offering a clear perspective on what it means to lead in an era where AI will reshape every workflow, every role, and every strategic decision. It encourages managers to stay grounded in their mission, prioritize human impact, and treat AI as a tool to amplify—not replace—human judgment and expertise. It closes with a call to action: to keep experimenting, keep learning, and keep building the organizational muscles that make AI both powerful and safe. With the frameworks, tools, and playbooks from this book, you now have everything needed to guide your teams through the next phase of AI transformation—one that is thoughtful, responsible, and built to last

Leading AI
Transformation
AI
Scaling
& reuse
Responsible
& safe AI
Use-case
selection
Architectures
& roadmaps
Literacy &
foundations

19 About the Authors

Brittanee Charles is a renowned systems transformation specialist. As a Fractional Business Systems Leader, she specializes in rescuing overwhelmed technology teams from the chaos of reactive operations. Her mission is to transform your systems from bottlenecks into strategic accelerators. Most systems teams are trapped in reactive mode. As if managing. GTM systems aren't chaotic enough; they're drowning in requests, juggling disconnected tools, and struggling to meet expectations. She assists by stepping in as a fractional leader and strategic advisor to create the structure, processes, and frameworks you need to thrive. With a proven track record across scaling startups and enterprise environments, she helps organizations reimagine their technological potential through people-centered, scalable frameworks.

Claude Louis-Charles, PhD, is a growth-driven technology leader with over 30 years of enterprise modernization experience, including business growth, team leadership, business partner collaboration, technical strategy development, and designing, building, and integrating complex cloud and AI solutions. Committed to delivering large-scale systems on time and on budget while building strong trust relationships. Dr. Claude has presented worldwide on IT challenges/solutions, including the World Health Organization in Geneva, Switzerland, the Ministry of Health in Uganda, the Department of Defense in Japan, and other major enterprise organizations.

Dr. Claude has multiple degrees, including a PhD in Information Technology with a specialization in Cybersecurity, an MBA, and a Certificate in Artificial Intelligence from MIT.

Made in the USA
Monee, IL
12 November 2026

8e92c3f3-0743-4dd0-9116-04fcbf9fe7cbR01